DON SIMBORG

THE FOURTH GREAT TRANSFORMATION

CREATING A NEW HUMAN SPECIES WITH AI AND GENETIC ENGINEERING

Published by
LID Publishing Limited
The Record Hall, Studio 304,
16-16a Baldwins Gardens,
London EC1N 7RJ, UK

info@lidpublishing.com
www.lidpublishing.com

A member of:
BPR
businesspublishersroundtable.com

Printed in the United States
ISBN: 978-1-912555-72-7
ISBN: 978-1-911671-29-9 (ebook)

Cover and page design: Caroline Li

DON SIMBORG

THE FOURTH GREAT TRANSFORMATION

CREATING A NEW HUMAN SPECIES WITH AI AND GENETIC ENGINEERING

MADRID | MEXICO CITY | LONDON
NEW YORK | BUENOS AIRES
BOGOTA | SHANGHAI | NEW DELHI

To Maddy

CONTENTS

PREFACE

"If we could replay the game of life again and again . . .
the vast majority of replays would never produce
(on the finite scale of a planet's lifetime) a creature
with self-consciousness."

—**STEPHEN JAY GOULD**, *Full House*

We, *Homo sapiens*, are the only human species left on Earth. For 98% of our time on the planet, though, other human species co-existed with us. Why have we survived while all the others have perished? And, specifically, why did the Neanderthals, our closest relatives, go extinct? They had brains slightly larger than our own, used tools, had some type of language, created wall art, buried their dead and had their own culture. The debate still rages as to whether we outsmarted them somehow, with better tools and language, or perhaps we just outnumbered them over time, with our continual migrations out of Africa. The evidence is simply insufficient to reach a definitive conclusion.

You may then wonder how we can predict our future, if our ability to nail down the past is so difficult. There was no written documentation available 39,000 years ago, when the Neanderthals perished. We have to rely primarily on a sparse fossil record and other clues from their dwellings to form a picture of them. In preparation for my previous book, *What Comes After Homo Sapiens?*, I researched the most likely paths to the next human species. While my conclusions are speculative, our ability to look ahead is on much firmer ground than our tools to study the past.

The ways we look at both the past and future have changed. We now have detailed, written documentation of our more recent history, including a comprehensive look at the most important biological evidence: our genome. Amazingly, we have also been able to coax out enough DNA from the fossils of Neanderthals and other extinct species to recreate their genomes. That has given us a tremendous boost in understanding our past and has opened up a new field of study called *paleogenomics*. Still, we are limited to the study of the genomes from only a handful of individual Neanderthals and other extinct humans, compared to our current databases of millions of individuals. So, many mysteries regarding our past remain. That said, however, our vast current data sources, combined with our computational capabilities, now enable us to make a more accurate prediction of our future trajectory.

There is much in today's popular science literature that envisions possible futures for humanity. Books like *Homo Deus* by Yuval Noah Harari, *Cosmosapiens* by John Hands, and *Homo Evolutis* by Juan Enriquez and Steve Gullans, all provide projections of the evolution of *Homo sapiens* over the next couple of centuries. The focus of these books, however, is mostly cultural and broad-based. Much of their focus is on the social and political aspects affected by our sciences and institutions, and paints a picture of only what *Homo sapiens* might be like in the future. None deal with the more specific scientific questions regarding speciation: exactly when and how a new species will arise from *Homo sapiens* in the course of evolution. Evolution of *Homo sapiens* is a description of how we change over time. Speciation is how *Homo sapiens* transitions to a totally new human species. This book focuses on speciation.

My 2017 book, *What Comes After Homo Sapiens?*, served to document my research on this subject. Readers were most interested in the science and evidence regarding the answer to the book's title question. *The Fourth Great Transformation* has been updated to the present and tells this non-fiction story in a broader popular-science manner, albeit one with a speculative ending.

The Fourth Great Transformation ends with the emergence of a new human species as a result of our using our two most advanced tools: artificial intelligence (AI) and genetic engineering. I will describe in detail one possible scenario as to how this could occur. It is only one potential scenario out of many that could have been chosen.

Each of the first three great transformations deserves a book in itself. All are equally momentous in the history of life on Earth leading up to this *Fourth Great Transformation*, but they will be described only briefly here.

The Earth has existed for about 4.5 billion years. The *First Great Transformation* was the emergence of life on this planet about 3.8 billion years ago. Exactly what this first life form was is unknown to us, as is where on Earth it occurred or even whether it originated on Earth or was seeded from elsewhere in the cosmos. The only evidence of these original life forms in existence today are the ubiquitous single-celled organisms called bacteria and archaea,[1] which together are called prokaryotes. All other life forms – including us – evolved from these entities.

After those first single-celled prokaryotes appeared on Earth, not much happened with them for the next two billion years. Think about that. Then suddenly (in evolutionary terms), a new life form took over about 1.8 billion years ago. This happened when a species of archaea engulfed a species of bacteria, and those prokaryote cells transitioned into the eukaryote cell. The most dramatic difference between a prokaryote and a eukaryote was that eukaryote cells now specialized into team players to become multi-cellular organisms. Some cells become heart cells, others skin cells, leaf cells, root cells and so forth, specializing into all the various cells that make up plants and animals.

Once the forces of evolution were unleashed by this *Second Great Transformation*, countless variations in life occurred and led to billions of new species of plants and animals. Only a small fraction of these species remains alive today. Most of that evolution occurred over 1.5 billion years and long before anything like a human appeared on Earth. The immediate precursors to humans date back to only about 7 million years ago. It is still being debated exactly who were the first humans – possibly *Homo habilis* or *Homo ergaster.* They did not arrive until about two million years ago. But, when they did, it ushered in the *Third Great Transformation.*

That *Third Great Transformation* was the development of our human brain, which began in earnest at about that point two million years ago. This transformation from the relatively primitive, ape-like brain to our amazing and mysteriously complex neocortex has brought us intelligence, consciousness, language and tool-making capabilities unparalleled in nature.

This *Third Great Transformation* ultimately enabled us to create two particularly remarkable tools – AI and genetic engineering. It is these tools, used in combination, that will generate the F*ourth Great Transformation.*

THE FOURTH GREAT TRANSFORMATION

The first three great transformations are fact. The fourth is speculation, albeit based on hard scientific evidence and our rapidly accelerating tools capabilities. The disciplines of evolutionary biology, taxonomy, species and speciation, genetics and genomics, AI, neuroscience, nanotechnology and genetic engineering (among others) all inform the study of our future evolution as a species.

A study of the literature of those disciplines and consultation with experts in those fields led me to the following hypotheses:

1. Another human species, let's call it *Homo nouveau* ('new man'), will co-exist with us within the next two centuries.
2. *Homo nouveau* will not emerge naturally in the course of evolution, as we *Homo sapiens* ('wise man'), did. Instead, it will be the result of our tool use.
3. Those tools will be AI and genetic engineering.

If these hypotheses prove true, they will represent the *Fourth Great Transformation* in evolution.

We are already going through the preliminary phases of genetically altering thousands of other species. In the short period of several decades, we have come to the point where GMOs (genetically modified organisms) have become a routine part of agriculture and our food supply. They have become commonplace with regard to plants, and in the past few years have extended to animal foods like salmon and to so-called *pharm-animals*, genetically modified to produce drugs for human consumption.

The extension to GMHs (genetically modified humans) has also begun, with our attempts to cure genetic diseases and even cancer. It is only a matter of time before we not only alter species, but also perform genetic manipulations that lead to entirely new ones, including next-generation humans. This won't happen suddenly. The more we learn about our genome, the more complicated we find it. And, although tools like CRISPR and other genetic engineering methodologies we'll examine here are being used more universally, we remain a long way from being able to apply them safely and confidently in most situations.

This is where our other advanced tool, AI, comes in. We will use AI to overcome the complexities and problems in genetic engineering. This will become clear in the following chapters.

The first three great transformations have all been sequential, with each subsequent one depending on the previous. The *Fourth Great Transformation*, however, will be different; it will be caused and created entirely by the collective brains of *Homo sapiens*.

An important question should be asked here: when *Homo nouveau* arrives, will that truly rise to the level of a great transformation? The first three great transformations were each created by myriad random natural events over very long periods of time. Each was dramatic, mysterious and highly unlikely. Each of their probabilities was so remote that the probability that *all three* have occurred on Earth (or anywhere) seems extremely small. Even our own existence was unlikely. As the late paleontologist Steven Jay Gould said, "If we could replay the game of life again and again ... the vast majority of replays would never produce (on the finite scale of a planet's lifetime) a creature with self-consciousness."[2] And yet, here we are, on the cusp of a *Fourth Great Transformation*.

But this one will be different in a rather striking way – when *Homo nouveau* emerges, we will be aware of it. In the entire history of life on Earth, there has never been a species that was aware that another species has evolved from it. In fact, there has never been another species that even had a concept of species. We, *Homo sapiens*, made up that concept. It is a figment of what we call our *fictive* thinking – our unique ability to think and talk about concepts that have no objective reality.

Fictive thinking allows us to think symbolically about things in ways that enable us to plan ahead and cooperate to an extent far more effectively than any other species.

The first known discussion of the concept of species is from the ancient Greeks. That's only a couple of thousand years ago – long after any other human species was alive. It is doubtful that any other human species thought about this, let alone recognized the transition to another human species.

More importantly, the *Fourth Great Transformation* will mean that we have transformed evolution itself. The true greatness of this transformation will be that evolution in the human lineage will never be the same as in the past. Although Darwinian natural selection will continue, our genetic manipulations will usurp the normal and natural evolution that has led to new human species in the past.

And we will continue to preempt those human speciation processes for as long into the future as humans exist.

CHAPTER 1

EVOLUTION AND SPECIATION

"The essence of the 'species problem' is the fact that, while many different authorities have very different ideas of what species are, there is no set of experiments or observations that can be imagined that can resolve which of these views is the right one."

—**JOHN BROOKFIELD**, Professor of Evolutionary Genetics, University of Nottingham

In order to understand how the *Fourth Great Transformation* will be so astonishingly different from anything in the past, it will be helpful to understand how evolution and speciation has occurred up to the present.

WHAT DOES IT MEAN TO EVOLVE?

The term *evolve* was used in the Preface of this book on the assumption everyone knows what that means. In truth, not everyone agrees on what it means. For millennia, up until the 19th century, there was a generally held belief that the amazingly beautiful and diverse panoply of plants and animals could only have been created by an intelligent designer. And, certainly, the designer created mankind in its image. Furthermore, not only did the designer create all of the species independently, but commanded that they were immutable. That is, they didn't evolve.

A brilliant man named Darwin, among others of his time, considered that perhaps species were not immutable, but had all evolved over time from "...one living filament, which THE GREAT FIRST CAUSE endued with animality, with the power of acquiring new parts, attended with new propensities..."[3] In other words, Darwin had a concept of evolution of all species over time from some original common ancestor. He further had the notion that only the 'fittest' were able to survive among competing organisms. This man was **not** the Charles Darwin who wrote *On the Origin of Species by Means of Natural Selection* in 1859, but Erasmus Darwin (his grandfather) who lived in the 1700s.

A number of naturalists and biologists had begun to think in terms of evolution rather than creation by the early- to mid-1800s. In 1809, the famous French biologist and taxonomist Jean-Baptiste Lamarck published *Zoological Philosophy: An Exposition with Regard to the Natural History of Animals.* He outlined one of the first theories of evolution based on the biological principle that animals acquired inheritable characteristics due to the influence of the environment. In particular, he stated that either excessive use or disuse of body structures changed them in ways that would be passed on to their offspring. The frequently used example of such Lamarckian evolution is that the giraffe developed its long neck by continuous stretching to reach high tree branches for leaves. Each stretched generation passed on this elongating neck to the next generation. Lamarck also believed that inheritance was a blending of the characteristics of the parents into the offspring. Although his basic theories of evolution and inheritance have been debunked, some elements of Lamarckian evolution have reappeared in our understanding of the *epigenome* – the set of chemical processes that affect the expression of genes. We'll talk more about that later.

By the time of Charles Darwin's epic publication in 1859, another noted British scientist, Alfred Russel Wallace, had developed a theory very similar to his that involved natural selection as the principle driving force of evolution. Although Darwin had been working on this idea for decades, it became apparent that Wallace might publish the theory ahead of him.

Instead, they agreed to publish their theories and present them together. Darwin generally gets most of the credit because he researched and published on the subject far more extensively than Wallace. Nonetheless, they should be considered co-creators of the most important and most enduring evolution theory, which relegated creationism to a much smaller religious following, which persists today.

DARWINIAN EVOLUTION

A number of concepts in evolution by natural selection distinguish it from creationism. First, and foremost, is that there is evolution of species – that they are not immutable.[4] They change. Further, these changes are gradual.[5] Ultimately, over thousands of generations and millions of years, the changes become prominent enough that new species emerge from old ones. This is called *speciation*. Implicit in speciation is the notion of common ancestry of related species and, if one goes back far enough, that there is common ancestry among related organisms – ultimately, a single common ancestor of all living things.

The key to Darwinian evolution is natural selection, which requires two factors: 1) organisms must change; 2) those changes interact with the environment, and those that enable the organism to best adapt will persist.

With regard to the first requirement, neither Darwin nor Wallace knew anything about genetics, DNA or the basics of molecular biology. Although the Augustinian Monk Gregor Mendel was doing his pioneering work on genetics during those times, they appear not to have been aware of his work.[6] In fact, Mendel's work was not widely known and appreciated until the early 1900s.

What Darwin and Wallace did know is that the offspring of animals varied from each other and often from their parents. Further, animals in nature produced far more offspring than would survive into the next generation. Although they didn't know why different offspring had different characteristics from their siblings and/or parents, they did know that some characteristics made it more likely that some offspring would live and others would die. That is, some of the variations made offspring more adapted to the environment and thus better able to survive and reproduce. In essence, the environment 'selected' some offspring to survive – and later pass on their characteristics to future offspring – while others were selected to die before reaching the stage of reproduction. Over generations, the good variations would become prevalent in a species' gene pool while the bad ones disappeared. This process of natural selection drove evolution and ultimately speciation. Sometimes this process is called *survival of the fittest*. What fittest means is simply being best able to survive long enough to have more offspring than others.

TODAY'S VIEW

We now know a lot more than Charles Darwin knew, and have become much more precise in defining natural selection without changing its fundamental principles. For one, we know that the traits, structures and functions of all plants and animals are controlled by their genes. Darwin didn't know about genes. Genes consist of a chemical called DNA (deoxyribonucleic acid), which is stored in structures called chromosomes in the nucleus of every cell of a plant or animal. All DNA – and therefore all genes – consist of a long string of four kinds of building blocks called nucleotides, given the labels A, T, C and G, for the first letters of their chemical names.

Each gene produces one or more proteins that are required by the plant or animal for specific purposes, such as chemical reactions or physical structures. Those physical structures in humans include all of our organs, as well as the various cell types that make up each organ, and how these cells function and what chemicals they produce. Basically, proteins do all of the important things we require to be who we are, what we look like, and how we function in and adapt to our environment.

Proteins are made up of 20 different kinds of building blocks, called amino acids. It is the sequence of those As, Ts, Cs and Gs (the nucleotides) in the genes that determines which amino acids go into each protein, how many of each there are and in what order they are used to build the protein. The structures and shapes of these proteins determine virtually everything about a plant or animal.

It is important to understand that the proteins in all plants and animals consist of the same 20 amino acids, which are defined in the same way by the same four nucleotides in their genes, which are part of their DNA. The differences between a worm, a tree and a human are defined solely by the *sequence* of the nucleotides in our DNA.

Every cell in our bodies contains the exact same DNA, and therefore the same genes.[7] For example, the DNA in a heart cell has the same genes as in a pancreas cell. Clearly, the heart does not produce insulin, which is made only by our pancreas, and our pancreas doesn't pump blood around our bodies. It turns out that the genes that produce proteins constitute only about 1.5% of our DNA. We call our entire DNA our genome, but only this small percentage of our genome produces those proteins.

The rest of the genome was formerly thought to be junk. It turns out that a lot of that junk does a lot of important things. The most important thing is that some of it generates special chemicals that directly regulate the *expression* of genes at different times and in different cells in the body. Expression means whether that gene is active in producing its protein. This part of our DNA that controls gene expression is called the *epigenome.* The epigenome is key to how the body develops while in the womb and throughout life, and how it functions in the adult. It determines which genes are active in each cell and at what times during development or later they are active.

The reason a heart cell doesn't produce insulin is that the insulin-producing genes are not expressed in heart cells. Likewise, heart muscle cells are not expressed in pancreas cells. Therefore, the difference between a heart cell and a pancreas cell is controlled by the epigenome, not their protein-defining genes. The genes are the same. How this all works is simply amazing, complex, and to this day not completely understood. As we will see later, it is also the epigenome that determines many of the differences between species that, to a large extent, have the same genes.

Every living thing has a genome. The genomes vary from species to species in the number of nucleotides, chromosomes and genes each possess. There is nothing particularly unique about us in that regard. We have roughly the same number of genes as a mouse or a worm. The number of chromosomes does not tell us much about a species – for instance, we have fewer chromosomes than a pigeon. We have fewer total nucleotides, and therefore a smaller genome, than many so-called lower animals. And, as stated earlier, those nucleotides are the same in all species. It is only our nucleotide *sequence* that varies. Some of our genes are unique to humans, even though most are not, but what makes us special is how we *express* those genes.

So, let's get back to natural selection and evolution. The first requirement is that organisms must change. This means a change in the DNA sequence from one generation to the next – from parent to child. There are several major ways that this variation occurs in humans and other sexually reproducing organisms. The first is by mutation. A mutation is a substitution of one nucleotide for another.

These occur spontaneously and randomly in every organism. Some things make the rate of mutation higher, such as radiation, but they never stop happening in our bodies.

What the substitution is, and where in the genome the mutation occurs, determines what effect it will have. For example, if the mutation changes a nucleotide in a part of the genome that isn't in a gene or the epigenome[8] it has no impact on anything. Even if it does change a nucleotide in a gene or the epigenome, it still may or may not have an impact. A mutation, even in a gene, does not necessarily alter the amino acids in the protein that gene generates. That is because there is some redundancy in the code that determines which amino acid is generated from a given nucleotide sequence. More than one sequence can lead to the same amino acid.

Often, a change in a single amino acid in a protein doesn't alter its function. But sometimes it does. Proteins are complicated chemicals that curl up into a variety of shapes. If a change in the amino acid sequence alters its shape, it is likely that it would also alter its function. For example, a specific mutation in a single nucleotide in the gene for a hemoglobin protein leads to an abnormal form that causes the debilitating disease sickle cell anemia. Likewise, if the mutation is in an area of the epigenome, it could impact the gene's expression, which could be equally significant.

Our bodies have trillions of cells. Mutations don't occur everywhere in a body at the same time, but rather at random times, in random cells and in random places in the DNA. We have two kinds of cells in our bodies: *somatic cells* and *germline cells*. Germline cells are those that are involved in reproduction, like sperm in males and eggs in females. Mutations in those cells are heritable; they get passed on to offspring. These mutations are the most important, since they impact the next generation of individuals and collectively could alter the evolution of the species.

Mutations in the rest of the body – the *somatic cells* – are not passed on to offspring and only impact that individual. They could still be important, though, since such mutations are involved in the development of cancer. It is estimated that an average of about 60 germline cell mutations are passed on to each child from human parents. More of these come through the sperm than the egg. Since we have more than 6 billion nucleotides in our genome, 60 mutations doesn't sound very significant. But with more than 380,000 children born each day worldwide, the potential for introducing evolutionary changes is very real.

Besides mutations, there is a second major way that genetic changes between parents and offspring occur in sexually reproducing animals like us. In preparing the sperm or the eggs, only half of each parent's genes and other DNA is copied. The half that makes it is not solely from either the maternal or paternal side of each chromosome, but rather is an unpredictable mixture of each. This mixing creates new combinations of genes and epigenetic DNA that also could change, for better or worse, the adaptability of the offspring to the environment. That's why no two sperm or eggs are alike, and thus children of the same parents can vary dramatically.

Finally, there is a third – and perhaps the most significant – way that genome changes can happen between parents and offspring. In the process of copying the DNA into a sperm or egg, a variety of errors can occur. Entire genes can be omitted or copied multiple times, and entire chromosomes could be omitted or duplicated. In fact, as the evolutionary biologist Neil Shubin describes in his book, *Some Assembly Required*, duplication of genes – sometimes dozens or even hundreds of times in a genome – is an important mechanism of evolution because it is so common. This allows the organisms to retain the functions of the original copy, while mutations in other copies allow new adaptations to occur without destroying those existing functions.

Once a DNA change has occurred, natural selection requires that it has to impact the adaptability of the organism to the environment. That is, it must either make individuals in future generations more likely to survive to have more children or more likely to die before having children.

PROTEINS DON'T BLEND

As mentioned earlier, Darwin likely was unaware of the work of Gregor Mendel, who did the pioneering work on genetic inheritance. Mendel was an Augustinian monk whose studies of peas and other plants in the 1800s revolutionized our understanding of how inheritance works. Unfortunately, he worked in relative obscurity in his abbey in what was then part of Hungary. Furthermore, he published his work in an equally obscure German-language journal. So, the significance of his work went largely unrecognized until the early 20th century.

Mendel demonstrated that inheritance is not a blending of parental characteristics, but rather follows rules of dominant and recessive inheritance of discreet hereditary units. Long after Mendel died, these units were named *genes*. Since each person has two copies of each gene – one from each parent – if one or both copies of the gene are in the dominant form, then that controls the manifestation of the trait in the individual. Only if both gene copies are recessive would the recessive genes then determine the manifestation of the trait. This manifestation is called its *phenotype*, and the actual genetic makeup is called the *genotype*. Note that the phenotype is the same whether the child inherits a dominant gene from both parents or a dominant gene from one parent and a recessive gene from the other.[9] There is no blending of the two.[10] Furthermore, the recessive gene can still be passed on to the next generation. It does not disappear or get diluted in any way.

Darwin didn't know any of this. He still believed the common notion of his time that inheritance was a blending of parental traits. It was as though the inheritance factors were like paints going into a bucket where they would mix together. Had he been a mathematician rather than a naturalist, Darwin could have figured out that the blending theory would be inconsistent with natural selection. New mutations leading to new traits would be quickly diluted over a few generations, such that natural selection would not have the time or power to work.[11] Fortunately, his lack of understanding of genetics does not reduce the value of his evolutionary theory.

We now know what neither Darwin nor Mendel knew: that genes produce proteins. What dominance really means is that if a person has only one copy of a dominant gene, then that enables the person to produce enough of the protein necessary for whatever manifestation (phenotype) that protein creates. If the person has two copies of the dominant form, it may produce even more of that protein, but that does not change the phenotype. The recessive form of a gene either produces no protein or a different version of it that does not interfere with the function of the dominant protein or cause any other harm. In any case, the dominant protein and the recessive protein don't blend.

Then how do we explain the fact that some inheritable traits show obvious phenotype blending of parental phenotypes in the children? Skin color and height are two examples of this. It took the work of three famous statisticians in the early 1900s to reconcile these observations with natural selection and Mendelian genetics. They were Ronald A. Fischer, J.B.S. Haldane and Sewall Wright, who established the field of population genetics. They determined that such blended traits, or phenotypes, were the result of the interactions of multiple genes – genotypes – that impact the same trait. For example, we now know that there are hundreds of genes that play a role in determining the height of an individual. Fisher, Haldane and Wright worked out the mathematics underlying the compatibility of natural selection with Mendelian genetics.

It is easy to imagine how mutations or genetic errors can have a negative impact on an offspring. They can either deny the offspring an important protein, or they can generate a harmful new protein. All known genetic diseases, and there are thousands of them in humans, are caused by mutations or genetic errors that have entered the gene pool and continue to be passed on. Most of these eventually disappear because the offspring with them die before they have children or have a dramatically reduced reproduction rate.

However, we do know that some genetic diseases persist for long periods of time. Why is that? Why doesn't natural selection cause them to die out? Remember that a child inherits one copy of each gene from each parent. In some genetic diseases it takes only one copy of an abnormal gene to cause the disease. This means that the protein produced from this abnormal gene form is somehow harmful. Fortunately, those genetic diseases are rare, since the offspring are less likely to live to reproductive age. Sometimes, though, the negative consequences of a harmful gene don't appear until later in life, giving the individual ample opportunity to pass on that gene. An example of this is Huntington's disease, an inherited condition that causes nerve cells in the brain to break down over time. With modern medicine, many children with serious genetic disorders are now more likely to live well into reproductive age, which increases the likelihood of passing on harmful genes.

Many genetic diseases require the child to inherit the abnormal gene from both parents. That is, the gene is recessive, and the normal protein is absent. When a child has only one copy of this abnormal gene, we call it the *carrier state*. Since this person also has a normal gene copy, the normal protein is produced in sufficient quantity to prevent harm.

So, the child is normal, but carries the abnormal gene and could pass it on.

The carrier state does not confer any negative survival or reproductive impact, so it is less likely to die out in the population. That is because the probability of inheriting the abnormal gene from both parents is relatively small. In fact, many of these so-called recessive diseases actually provide some positive effect. In these cases, an individual's gene in the carrier state makes it more likely to permeate a population's gene pool. The gene for sickle cell anemia is an example. Individuals in the carrier state who function normally have one normal copy of the gene and are less likely to get malaria than people with two normal copies. Therefore, the environment selects positively for individuals with one copy of the sickle cell gene in areas where malaria is endemic. That is why sickle cell disease is still with us. It is also why it is more prevalent in African Americans, as malaria is more prevalent in Africa.

MOST SPECIES BECOME EXTINCT

It is one thing for genetic changes to affect the survival of individual offspring. It is quite another for the *collective* genetic changes in a species to affect the survival of the species or its evolution into another species (speciation). Evolution has produced billions of species, but more than 99% have become extinct for various reasons. Natural selection enables species to evolve in ways that make that species better adapted to its environment. What natural selection cannot anticipate is that the environment may change in ways not suited for those previous adaptations, or in ways faster than evolution can continue to adapt to. For example, the climate could change in such a way that a previously well adapted species can no longer survive in a new ice age or hotter environment. The food supply of a species can run out. Catastrophes such as volcanic eruptions or asteroid impacts dramatically alter the environment. New diseases could evolve and attack a species. Competitors could enter the environment that either kill a species, destroy its habitat or use up its food supply. That is certainly the impact that *Homo sapiens* has had on other species. The historical extinction rate of all species has increased 1,000-fold since we've arrived. As you can see, the list of environmental threats to species is quite long.

Natural selection cannot foresee environmental change. It does not foresee anything. The genetic changes are random. Natural selection will improve survivability in a given environment, but if that environment changes, those adaptations may not work as well. By chance, some gene pools will perform better in a changing environment than others. Or, some gene pools will be better able to adapt over time to the changing environment if the environment change occurs slowly enough. Some environmental changes, such as an asteroid impact or volcanic eruptions, happen too quickly for natural selection save a species. That seems to have been the case in most of the five mass extinctions that have already occurred on Earth.[12] The fact is, the vast majority of species that ever lived have gone extinct and new species have replaced them.

If an organism needs a better kidney or digestive enzyme, or needs to run faster to avoid predation, nothing in Darwinian natural selection causes any of that to happen in any greater likelihood. If it happens by chance, great. That change will then be selected by the environment and the organism will be more likely to survive. Note that the changes may not be optimal, as compared to what some intelligent designer might do. Thus, improvements through random genetic changes and natural selection are usually imperfect. There is no goal of perfection – just improvement in reproductive capability.

Another limitation or feature of Darwinian evolution is that the changes are small, incremental and limited by the existing traits of any organism. Millions of years ago, when animals emerged from the oceans onto land, they didn't suddenly, spontaneously develop feet through random genetic changes. Some fish gradually modified their existing fins to move about on land over thousands of generations. Even though some fish species evolved fins into legs, they also had to be able to breathe air through lungs rather than gills. It is unlikely that legs and lungs both appeared through random mutations at the same time. In fact, some fish had already developed lung-like organs in the water prior to their developing legs. Likewise, when bats developed wings, these new appendages didn't suddenly appear through random genetic changes. The existing forelimbs of an ancient ancestor gradually morphed into wings over many generations. It could be said that God produces angels by design, but Darwinian evolution produces bats through random genetic changes over long periods of time.

The scientific term for this is *exaptation* – the evolutionary adaptation of an existing structure or trait for some new purpose. For example, feathers originally served an insulation purpose in reptiles and later were used for flight in birds. The three middle ear bones we need for hearing – the incus, malleus and stapes – were originally jawbones in reptiles and had nothing to do with hearing. Billions of small transformations stemming from from random events got us here, not billions of new ideas.

As we will explore in Chapter 6, on genetic engineering, we are getting pretty close to being able to modify our genetic code in almost any way we wish. Yet, even when we achieve that, we will still not know how to make a better kidney. Our problem is not having a good typewriter; it is knowing what to type. *Homo sapiens* will not become an intelligent designer even though we will create *Homo nouveau*. There will be no *Homo deus*.

Instead, there will be a new kind of speciation event that will differ from all previous speciation events in evolution. Speciation requires special conditions in the course of evolution. To understand how the creation of *Homo nouveau* will be a different kind of speciation event, we will review how speciation has occurred up until now.

THE SPECIES PROBLEM

Evolution of a species does not necessarily mean it will evolve into a new species. Over thousands of generations and millions of years, a species can change substantially and yet still remain the same species. Special conditions must exist for these changes to result in a new species.

First, it is important to understand that there is no universally agreed upon definition of the word *species*. That is, we have what is known to taxonomists (people who classify things) as *the species problem*. There are many definitions, requiring us to decide which one to use in order to clearly describe when *Homo nouveau* appears. Let's start with what we know.

Our species is *Homo sapiens*. What does the term *Homo sapiens* mean? The obvious answer is that it refers to all of us human beings alive today and our immediate ancestors. But that answer is not so obvious and simple.

Homo sapiens is a Latin phrase consisting of two terms: *Homo* meaning 'man' and *sapiens* meaning 'wise.' Why Latin? The best answer is that it's a convention agreed upon by taxonomists for at least the past couple of centuries. Any language could have been used. The Swedish botanist and physician Carl Linnaeus, who lived in the 18th century, was the originator of the classification system we use today for all living things. He used Latin, and we've just continued that tradition.

Why two terms rather than something simple like *humans*? The Linnaean system is a hierarchy of groupings that begins at its lowest level with *species*. Linnaeus himself never really defined the word species and based his original groupings on a variety of observable features such as anatomy, how they reproduced, how they breathed and other variables. The many definitions used today came later. Species that are closely related are grouped into the next higher level, called *genus*. The most closely related genuses are grouped into the next higher level, called *family*, and so on up through a total of eight increasingly larger groupings, to the highest level: *domain*. There are three domains of life: Bacteria, Archaea and Eukarya. Humans are in the Eukarya domain. Table 1 below shows the full classification of *Homo sapiens*.

TAXONOMIC LEVEL	*HOMO SAPIENS* DESIGNATION	EXAMPLES
Domain	Eukarya	All plants and animals, fungi, amebas
Kingdom	Animalia	All animals
Phylum	Chordata	Vertebrates (fish, amphibians, reptiles, birds, mammals), lancelets, tunicates
Class	Mammalia	Cows, dogs, rodents, bonobos, marsupials
Order	Primate	Monkeys, lemurs, baboons, apes
Family	Hominidae	Gorillas, chimpanzees, orangutans
Genus	*Homo*	Neanderthals, Denisovans, *Homo ergaster*, *Homo heidelbergensis*
Species	*sapiens*	Today's living humans

TABLE 1 – **TAXONOMY OF HOMO SAPIENS**

Although the lowest level is called species, to accurately identify any species, both the name of the species and the name of the genus must be used. In the case of *Homo sapiens*, the species name is *sapiens*, and the genus name is *Homo*. The species name alone does not always uniquely identify a species because the same species name could be used in multiple genuses. Therefore, every species is identified by the combination of its genus and species.

By definition, since *genus* represents a grouping of related species, each genus usually contains multiple species. In our case, the *Homo* genus not only contains *Homo sapiens* – the only living member of our genus – but also many extinct species, such as *Homo neanderthalensis* (Neanderthals), *Homo erectus*, *Homo habilis*, *Homo heidelbergensis* and others. Note also that the word *human* refers to our genus and not our species. The Neanderthals were humans, for example. *Homo nouveau*, if and when they arrive, will also be humans.

So much for the semantics. The more important consideration is how taxonomists determine the groupings. First, what criteria are used to distinguish one species from another? Next, what further criteria define relatedness, which allows them to be grouped into a genus? Ultimately, how are genuses then grouped into families, and so on up the hierarchy? That's where the problem is.

WHAT'S THE PROBLEM?

The discussion of *Homo sapiens* is a good place to start. There are more than 7 billion people alive today. We occupy virtually every livable land-based niche on Earth. Our variations in physical, mental and cultural characteristics are enormous. We include Hispanics, Caucasians, Africans, Asians, Inuits, Islanders, Aborigines, Native Americans, tall people, dwarves, people born with one kidney, people with six fingers on each hand, people with an extra Y chromosome, and combinations of any of the above – and countless other – variations. Our languages, rituals, activities, diets, dwellings, organizations, governments and religious beliefs are likewise tremendously variable.

Yet, we are all one species: *Homo sapiens*. On the other hand, consider animals like a grey wolf (*Canis lupus*) and an Alaskan Malamute dog (*Canis familiaris*).[13] They look mostly alike, act similarly, even occasionally breed together, yet are classified as two different species. Note that they are in the same genus, *Canis*, since they are closely related. But humans today vary in their appearance, behaviors, living environments and other observable characteristics much more than wolves vary from Alaskan Malamutes. Yet, all humans today are one species, while wolves are a different species from dogs.

It all comes down to how we define species and who defines them. And that's the first hint of a problem. Who defines them varies depending on the type of plant or animal, and how they are defined varies with a confusing mélange of biological, anatomical, ecological, genetic and other criteria. The overall problem is that species have no consensus definition, leaving us with the species problem. We *Homo sapiens* have been debating that definition since the time of Plato. Detailed biological factors about chemistry, microscopic appearance, cellular metabolism, and DNA, RNA and proteins were unknown to early taxonomists. These are all used today in defining species, but there is still no agreed-upon definition. We can now sequence the entire genome of every living organism, but even that doesn't solve the problem. The genomes are different between every single human on Earth. How much difference would there need to be to declare a separate species? In what part of the genome? There are no correct answers here.

There is no objective reality in nature called 'species.' We *Homo sapiens* made up that concept with our fictive thinking, as described in the Preface. There are no species in nature that we can find, study and then name. Species exist only in our minds, and that's what makes naming and defining them so difficult. Yes, the organisms exist, but how we classify them is up to us.

The following will illustrate the issue of objective reality versus fictive concept. Imagine that some intelligent being from outer space came to Earth. It would be able to observe the continents, the oceans, the islands within the oceans, the lakes within the continents, the mountains and the plains, and all of the other physical characteristics of Earth. These are all objective realities that exist independent of the observer. Two independent such observers would see the same things and be able to describe them in the same manner. But, they would not see the United States, Canada and Great Britain, nor Nebraska, Idaho and England. Countries and states are not objective realities. They are concepts made up by *Homo sapiens* and are not observable. Likewise, such observers could see animals and plants and, if they had the right tools, could see bacteria, archaea, fungi and other living organisms. Yet, they would not see species because these are analogous to countries and states.[14]

A dozen different methodologies are used today to define species.[15] But, these depend on who is doing the defining and what part of the organism spectrum is being defined. Some methodologies are based on anatomy, ecological niche and behaviors, DNA sequences, etc., but most are some combination of these and other factors.

THE NON-EXISTENT 'OFFICIAL' SPECIES LIST

Who decides whether a new fossil or a group of organisms is a new species? Who decides what to call it? Are there official lists of species? Yes, many of them, and no two lists are alike. Sometimes species are determined and named by committees of experts. For example, there is a committee of the American Ornithologists Union called the Committee on Classification and Nomenclature of North and Middle American Birds. Their official list of bird species changes every year. Sometimes the committee decides that one species needs to be split into two based on their breeding behaviors, geography or new genetic analyses. Sometimes two species are merged into one for the same reasons. There are even examples of the same species going in both directions at different times. One wonders whether the changes are due to the birds or the committee members. There are similar committees for amphibians, reptiles and other groups.

However, there is no such committee for humans.[16] This is particularly problematic for looking at extinct human species. As new fossils are discovered and described, the scientists who publish these discoveries usually attempt to make a case for a new human species based on anatomy. That was true for the earliest humans, such as *Homo ergaster* and *Homo erectus*. Then, commentary at professional meetings or subsequent publications debates whether to accept the new species designation or to just lump the new fossil into an existing species. A consensus may not be reached. In the rare case of *Homo denisova*, DNA analysis was the only criterion on which to declare a new species, since there was inadequate fossil evidence. But, DNA analysis alone does not solve the species problem. Even though we know the full genome of the

Neanderthals, there is still debate today over whether or not they are a different species from *Homo sapiens*.

SO HOW WILL WE KNOW?

Given the confusion over defining species, how will anyone make the claim that a new species, *Homo nouveau*, has arrived? What definition will be used?

I have found the most useful definition to be that of Kevin de Queiroz, an evolutionary biologist at the Smithsonian National Museum of Natural History. He has proposed that "a species is a segment of a separately evolving metapopulation lineage."[17] The key phrase here is 'separately evolving.' That is the one criterion that has broad agreement among taxonomists. To qualify as a species, the group (the *metapopulation*), however defined, must be evolving independently of all other species, even if they live in the same environment or ecological niche and even if they sometimes breed with other species. This implies that their gene pools are not continually mixing.

To put it another way, there is something that generally keeps the group from interbreeding with members of other species. Even if they do interbreed, the interbreeding is infrequent and/or they are not likely to produce viable offspring. The term for this is *reproductive independence*, and what maintains that independence is called a *reproductive isolation barrier.* The group will then become increasingly differentiated genetically from all other species over time. The reproductive isolation barrier does not have to be absolute. There can be some occasional interbreeding, like there was between *Homo sapiens* and extinct human species. But, that small amount of interbreeding does not prevent the species from continuing to evolve independently.

WHAT ARE REPRODUCTIVE ISOLATION BARRIERS?

Like everything else related to species, the science of reproductive isolation barriers is complex and somewhat murky.

There are two major types of such barriers in sexually reproductive animals like us: *pre-zygotic* and *post-zygotic*. (A zygote is an egg fertilized by a sperm). A pre-zygotic barrier prevents a male and a female from two different species from mating in the first place or, if they do mate, prevents development of a viable fertilized egg. The most common of these barriers is physical separation. For example, if the two different species live on different continents, they would never have the occasion to mate. That doesn't mean that they couldn't successfully mate if brought together, but during the time of their separation no interbreeding occurs. The presumption is that these two species would continue to evolve independently over time. The physical barrier doesn't need to be as large as separate continents. It could be across a river, or a mountain, or at different altitudes on the mountain, or even the difference between a treetop and a tree root.

Another type of pre-zygotic barrier is some kind of incompatibility between the sperm of one species and the egg of another. Even if sexual contact occurs, the sperm does not successfully fertilize the egg for some physiologic reason, such as immune system rejection, chemical incompatibility or genetic incompatibility.

When two species are separated, they tend to evolve differently. That's because new mutations are not likely to be the same in the separated groups. Also, the environmental factors for natural selection may be quite different in different areas. A common event leading to speciation, therefore, is when a subset of a population separates from the larger group and stays separated for a long period of time. This could happen because of migration, or being carried accidentally by wind or water currents, or any number of other mechanisms. Over time, the separated group could evolve in different directions and develop a barrier to interbreeding. This is likely the case for the early humans, as will be discussed in Chapter 2, because of their small numbers migrating into isolated areas. This also happened to other animals when separated by tectonic plate shifts that created new continents. Note that in these speciation events, the original species continues as the same species (and continues to evolve) while the separated group is designated a new species.

One final and interesting type of pre-zygotic barrier is called behavioral isolation. This is when the males and females are not attracted to each other because of some appearance or behavioral trait. In humans these traits could be cultural or religious. This barrier is sometimes called sexual isolation or sexual selection and can occur even when there is no physical separation.

The second major type of reproductive isolation barrier is called a post-zygotic barrier. Here the sperm from one species and the egg from another species do successfully form a viable fertilized egg. But, after that, something goes wrong. One possibility is that the egg does not implant properly into the uterus or is somehow rejected by the mother. Or, the egg does implant, and an offspring is born, but that newborn is sterile. An example of that is the mating of a male donkey with a female horse to produce a mule. Mules are sterile and do not reproduce. Another possibility is that a newborn is produced that is somehow mentally and/or physically unattractive and less likely to mate.

In Chapter 2, all of the human speciation events to be described were the result of a pre-zygotic reproductive isolation barrier. In Chapter 8, it will become clear that the creation of *Homo nouveau* will be because of a post-zygotic barrier.

CHAPTER 2

HOW DID WE GET HERE?

"In other words, you do not need to change very much of the genome to make a new species."

—**DR. KATHERINE POLLARD**, Director,
Gladstone Institute of Data Science and Biotechnology,
University of California, San Francisco

How we got here is a 3.8 billion-year-old story starting with the *First Great Transformation* – the emergence of single-celled life on Earth. Exactly how that happened remains a mystery. Even the definition of life is not agreed upon by all scientists. For example, most scientists don't classify viruses as living even though they share many of the same chemicals and characteristics of prokaryotes. What viruses lack is the ability to live independently of another organism, which is why they are not considered alive.[18]

What is also puzzling about the emergence of life is that it only happened once – 3.8 billion years ago – as far as we can tell from our study of the dating of DNA histories. New lines of prokaryotes have not arisen at later times; they all date back to and evolved from that original time. And no matter how hard we try, we have not been able to recreate life in a laboratory. This has raised the possibility that although conditions on Earth have been able to sustain life all these years, we may not have the necessary ingredients and conditions to create life. That is why some scientists believe life arose elsewhere in the universe and came here when an asteroid or other celestial object crashed into Earth.

It took another 2 billion years until prokaryotes evolved into a totally new form of life, called eukaryotes. A eukaryote cell differs from a prokaryote cell in that most of the DNA is now contained in a cell nucleus, with its own nuclear membrane within the cell. Prokaryotes do not have nuclei. In addition, the engulfed bacteria now remain as a group of small organs within the cell called *mitochondria.* Mitochondria provide the power for all of the cell's chemistry. Prokaryotes do not have mitochondria. These dramatic changes ushered in the *Second Great Transformation*, and these new eukaryote cells became the basis for the evolution of all subsequently more complex life on earth, including all plants and animals.

In addition to the ability of eukaryote cells to specialize and become part of multicellular organisms, some eukaryotes dramatically altered their mode of reproduction. They switched from asexual division through a process called *mitosis* to sexual reproduction through a more complicated process called *meiosis.*[19] However, the focus of this book is to convince you that the *Fourth Great Transformation* is coming soon (in evolutionary terms), and we only need to go back as far as the *Third Great Transformation* – the emergence of the human brain – to help understand that. That brings our focus to the most recent 15 million years.

PRE-HUMANS

Our closest *living* relatives are the chimpanzees and bonobos. Genomic analysis shows that 98.8% of our genes overlap with chimps, more than any other living species. Although that seems to be an astonishingly small amount of difference, in reality that 1.2% represents millions of nucleotide differences. And that is just in the genes. There is evidence from genomic studies that most of the differences between chimps and us is in the epigenome.[20]

Chimps are not our closest relatives, however. Those would be the extinct Neanderthals, whose genes overlap with ours by 99.7%. What does closeness really mean, and how did that happen in the course of Darwinian evolution?

The great apes include orangutans, gorillas, chimpanzees, bonobos and us. We are all in the taxonomic family *Hominidae*. The Hominidae family is lumped together with other families that include monkeys, into the next higher level of classification called *order*. We are all in the *primate* order. (See Chapter 1 for a table of our taxonomy.) About 14 million years ago, all Hominidae had a common ancestor with the other primates that are not in our family and split apart from them in the course of evolution. (See Figure 1.) We don't know exactly what that common ancestor was, but we do know from the fossil record that all of us lived in Africa. We Hominidae and the other primates have all evolved away from that common ancestor, which no longer exists.

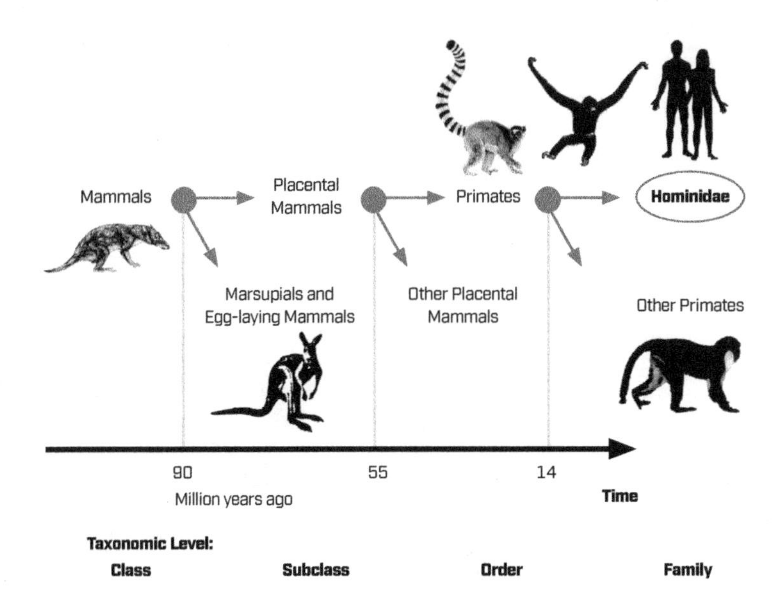

FIGURE 1 – **EVOLUTION OF MAMMALS**

Within Hominidae, orangutans split off onto their own evolutionary path about 14 million years ago, followed by gorillas about 7 million years ago. (See Figure 2.) Focus your attention, though, on the subsequent split between the lineage that led to the *Pan* genus (chimpanzees and bonobos) and the lineage that ultimately led to the *Homo* genus (humans). This occurred somewhere between 7 million and 5.4 million years ago. We chimps and humans had an unknown common ancestor in Africa before that split, but it likely looked more like today's chimps than anything human-like. It was probably a hairy creature that lived in trees and swung from branch to branch using its upper limbs. Its long arms and shoulder flexibility enabled easy tree navigation. In the unusual times when it was on the ground, it knuckle-walked on four legs and occasionally rose up on the two hind legs. It had both an opposing thumb on the forelimbs and an opposing toe on the hind limbs for easy tree navigation.

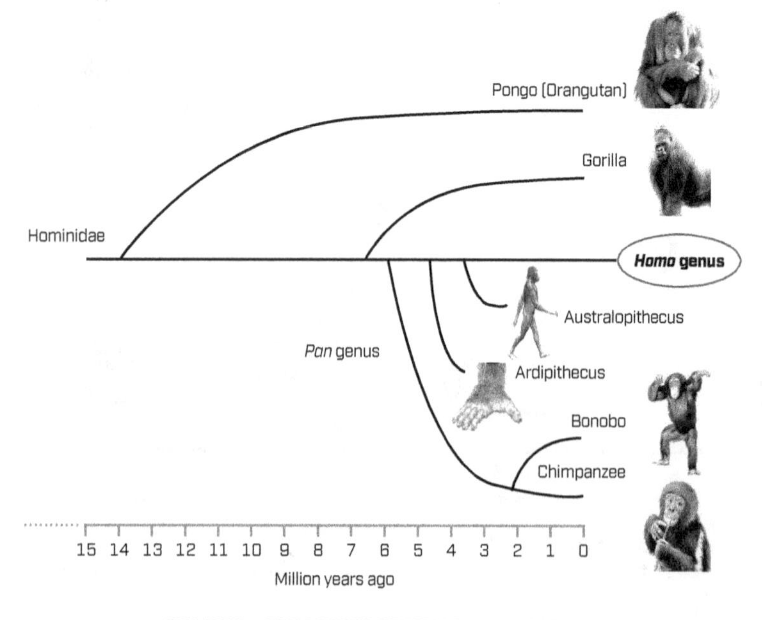

FIGURE 2 – **EVOLUTION OF HOMINIDAE FAMILY**

Like all animals, the chimpanzee line has evolved in the past 5–7 million years since they split from the common ancestor with humans. About 2 million years ago, the smaller and behaviorally different bonobo line branched away from chimpanzees as a separate species. However, the members of the *Pan* genus (which includes both chimpanzees and bonobos) still are hairy creatures that live in trees, swinging from branch to branch with long arms. Their brains are about the same size as they were 7 million years ago. On the other hand, the transformations on the lineage leading to humans have been much more dramatic.

Being the anthropomorphic creatures that we are, we have had much more interest in the human lineage and have studied it more intensely. Unfortunately, the fossil record is not rich enough to settle all debates about every step along the route to *Homo sapiens*. Clearly, somewhere along the line the *Third Great Transformation* occurred in the development of the *Homo sapiens* brain.

We will get to that in more detail, as it is the main reason we dominate on Earth in so many ways and chimps are in our zoos. There are other large differences besides brain development and all of its consequences. We rarely are in trees, walk bipedally upright on the ground and no longer have an opposable great toe. We lost most of our hair and we sweat a lot. We live in virtually every land-based habitat and we number in the billions, while chimpanzees are threatened with extinction in a shrinking habitat. At least for the time being, our line has had much more survival success than theirs.

So, imagine these two parallel lines of evolution for the past 5–7 million years from a common ancestor. The chimps changed a little bit and the humans changed a lot. Why? On the chimp side, they remained in the trees of Africa and hardly changed their lifestyle at all. On the side leading to humans, we grew brains three times the size of chimpanzees, became bipedal, lost our hair and wandered off to the rest of the world. How could Darwinian natural selection have treated us so differently?

Between our chimp-like ancestors and the first humans, we have found fossils of many different creatures that appear to be intermediate, between us and them – still ape-like, but less so. How less so? Looking at Figure 2, you will see the names of two groups of species: *Ardipithecus* and *Australopithecus*.[21]

Ardipithecus ramidus, nicknamed Ardi, is a fossil discovered in Ethiopia of a female who lived about 4.4 million years ago. Her brain was chimp-sized. Her feet looked like a chimp's in that they had an opposable big toe for tree climbing, but were also much flatter, indicating she could walk on them. Looking at her hands, she clearly did not knuckle walk, indicating that she was bipedal when walking. There were other skeletal features as well that were consistent with a more upright posture.

The most famous fossil of all time, however, was nicknamed Lucy. Lucy is also from Ethiopia and lived 3.2 million years ago. She is a member of the species *Australopithecus afarensis*. Lucy was unearthed before Ardi, and reports of her discovery were heralded around the world because she clearly was a bipedal ape-like creature that was touted as the missing link between chimps and us. Unlike Ardi, Lucy had no opposable big toe; her foot was human-like.

There has been a long-standing debate as to which came first: bipedal gait or a large brain. That is, did we walk upright before we got smart or vice-versa? Evidence from the *Australopithecus* period suggests the answer could be that they occurred during the same time period. The *Australopithecus* brain was about 20% bigger than a chimp's, still quite small by human standards – as noted, ours is three times the size of a chimp's. It appears, however, that *Australopithecus* was the first species ever to create a crude tool.

This implement was a sharp flake of rock created by knocking two larger rocks together. These flakes were clearly used to help butcher small animals.

It takes advanced brain activity to think ahead to what tool to make, how to craft it and how it will be used. It is one of the key characteristics of the *Homo* genus that came later. Thus, a small increase in brain size seems to have been accompanied by a major new brain function. This occurred at roughly the same time that our ancestors transitioned to an upright gait. In addition, there is evidence that the *Australopithecus* brain took longer to fully develop postnatally than other apes, which again seems to be a transition toward the very long childhood brain development period of humans.[22]

It is easy to imagine the natural selection advantage of a smarter brain. But why bipedal walking? Living and sleeping in trees clearly provided protection from many ground-based predators that prowled their territories in Africa. It is still unsafe for humans to walk the jungles of Africa. Why did we come out of the trees? There are many theories.

One theory was that as the forests of Africa began to thin out in some areas and became savannahs because of climate change, these early pre-humans had to travel on the ground to get from tree to tree. But Ardi and Lucy clearly lived in the forests, not the savannahs. Anyway, lots of animals walk long distances just fine on four legs. They didn't need to be upright to see over tall grasses since there weren't tall grasses at the times and places these fossils are from. They didn't need to free up their hands to hold spears or other weapons to kill game; they didn't have weapons.

To quote from my previous book, *What Comes After Homo Sapiens?*, one of the most accepted explanations is based on the work of a well-known anthropologist, C. Owen Lovejoy:

> "The explanation is somewhat complex and difficult to understand, but it goes something like this: bipedalism freed the *Australopithecus* mother to be able to hold onto one or two offspring while still becoming pregnant with another, thus helping overcome the limitation in numbers of children she could have. Great apes have a child only once every four or five years. Freeing of the mother's hands was needed because the infant hand and foot couldn't cling by itself to the mother like a chimp baby does. *Homo sapiens* is the only species today in which parents hold hands with their offspring while walking to protect them from all sorts of dangers – and they continue to do this long after the offspring are able to walk. Our ancestors probably did the same.
>
> Having free hands also enabled the *Australopithecus* father to carry back more high-energy food from his foraging outings. All of this led to longer term bonding between a male and female toward the evolution of the human nuclear family. After all, a male wouldn't risk going away for long periods of time to find food if his female didn't remain committed. I'm not sure I'm buying all of this, but it's one theory."

There are many other theories. Personally, I like the answer attributed to a Tibetan scholar (probably apocryphal) when asked why we are bipedal: "A sense of humor."[23]

So much for why we are bipedal. Why did we lose our hair? We can explain this one a little better. Eventually, pre-humans and early humans did leave the trees and became wanderers in the savannahs of Africa, and ultimately throughout the world. They were no longer in the shade of the trees all day. Can you imagine wandering around in the hot African sun in a fur coat? Natural selection would have much preferred a naked ape like us, as it would have allowed more cooling as the air moved across the now upright body. We know that we didn't actually lose our hair. The mutation just changed it from the thick fur that chimps have to the very fine, almost invisible, hair that humans have over most of the body. We also have evidence that humans evolved high densities of the type of sweat glands involved in cooling during this period. Chimps have far fewer of this type of sweat glands and don't sweat much.

In addition, chimps are light-skinned. Losing hair would have been cooler, but it would have exposed those creatures to dangerous solar radiation and sunburn. None of these pre-humans were smart enough to invent clothing or sunblock. We know that most of today's Africans have dark skin. We also know that there are many other humans in cooler northern climates who have light skin. There is a gene in the human genome labeled MC1R that generates the brownish pigment melanin, which causes dark skin. Melanin also protects the skin from damaging ultraviolet rays from the sun. All dark-skinned Africans today have the MC1R gene, but many other humans in cooler locales lack it. It seems likely that a mutation generated the MC1R gene back when the pre-humans and early humans left the trees and was easily selected by the hot environment. This increased their melanin production to protect them from the sun as their hair thinned out.

As humans migrated to colder climates, other mutations negated or reduced the effectiveness or expression of that gene and were later selected. That's because there would be some negative consequences to blocking too much of the sun's rays. We need ultraviolet light to produce vitamin D, which strengthens our bones. Having natural selection moderate the amount of melanin against the amount of sunlight seems like a reasonable compromise (not that reason has anything to do with mutations and natural selection). Those same mutations in the MC1R gene that reduce its effectiveness probably occurred in Africans over this time period, but these would not have persisted because the hot African environment would have selected them out.

Are Ardi and/or Lucy the direct ancestors of humans? Or, are *Ardipithecus* and *Australopithecus* evolutionary offshoots that led to extinction, while some other unknown pre-human led to the first humans? Figure 3 shows a couple of the theoretical options. It will take more fossils, or perhaps some ability to examine the DNA of older fossils, to sort that out. But surely there were missing links or intermediaries that were something like Ardi and Lucy.

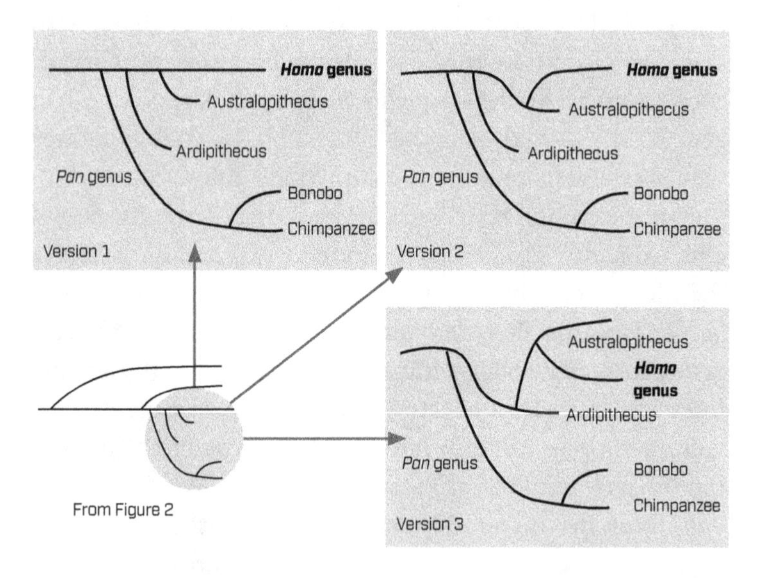

FIGURE 3 – **THREE POSSIBLE EVOLUTIONARY PATHS TO THE *HOMO* GENUS**

THE FIRST HUMANS

At some point the pre-human brain began to grow much faster. One would think that's when the pre-humans became humans. However, the problem is that there was no such point. The transition from pre-human to human is another one of those fictive ideas of *Homo sapiens*. Indeed, something magical happened between about 2 million years ago and 1.5 million years ago that set us on the path to the *Homo sapiens* brain. But, there is no way to pin down exactly what happened, and when and where, with precision. The *Third Great Transformation* happened over hundreds of thousands, and likely millions, of years in Africa, but that's about the best we can say. Except that it surely happened.

The first species that many classify as human is the 'handy man.' *Homo habilis* first appeared about 2.1 million years ago. This species clearly created and used those same sharp stone flakes for butchering small game as the Australopithecines before them. Their brains were larger than *Australopithecus*, and they had other morphological features that were closer to later humans.

It is important to distinguish the creation of tools from the use of tools. Our decision to classify something as *Homo* (human) is based in part on our assessment of advanced brain capability. Certainly chimpanzees, apes and virtually all other animals have brains that do many amazing things. Ants harvest fungi, for example. Birds build nests. Beavers construct dams. Spiders spin amazingly engineered webs. Honeybees communicate to their honeycomb mates the direction and distance to food sources by dancing. Examples of advanced brain function are everywhere in the animal kingdom.

We have settled upon tool creation as one of the defining characteristics that qualifies a species as being human. As we've said, tool creation is different from tool use. Other animals use tools. For example, chimpanzees use sticks to pull high-protein termites from their nests and use rocks to crack open nuts. Some monkeys also use rocks to crack nuts. Puffins, a kind of seabird, use sticks to scratch themselves, as do elephants. Crows can use sticks or other objects to get at food. Crows can even modify sticks into hooks, which some would argue is tool creation. Likewise, chimps have been observed selecting certain specially shaped branches and then stripping and trimming them to make a foraging tool. That also could be considered tool creation. However, as we will see in Chapter 3, humans have the market cornered on tool creation.

So, why would anyone consider *Homo habilis* the first human, rather than one of the *Australopithecus* species? Frankly, it is an arbitrary call – the transition was gradual, and the very definition of 'human' remains murky. Another species, *Homo ergaster* ('working man'), which lived 1.9–1.5 million years ago, could be considered the first human. *Homo ergaster*'s brain size was still larger and was intermediate in many features between *Homo habilis* and *Homo erectus*. They advanced on the stone flake idea to create a much more developed hand axe. They also used fire to cook both meat and plants in order to extract more nutrition, to feed their larger brains. Ian Tattersall, my favorite paleoanthropologist, is the curator emeritus at the American Museum of Natural History in New York City. He believes that *Homo ergaster* should be designated the first human,[24] so I will go with that. As I've said, it is an arbitrary call, but we will be more definitive later, when we discuss the creation of *Homo nouveau*.

There is no debate that *Homo erectus* was human. *Homo erectus* is the longest surviving human species to date, living from 1.8 million years ago to about 100,000 years ago. Its brain size was significantly larger than *Homo ergaster*, although smaller than later humans. It used more advanced tools and had control of fire for warmth and cooking. Since they were a fully ground-based creature, sleeping around a fire probably also provided a level of protection. They were able to build some kind of sea craft and to navigate on the ocean, which got them to various islands only accessible by boats or rafts. They also appear to have planned settlements where food preparation areas were distinct from living areas. Many believe they could not have done all of this without some kind of language.

Clearly, by 1.5 million years ago in Africa, we were well on our way toward the *Homo sapiens* brain. Darwinian theory requires that there must have been some genetic mutational changes in the pre-humans and early humans that led to better adaptability to survive in the environment. Without having the genetic analysis available of species that old, we can get some clues from genomes of living and more recently extinct species.[25] However, these are just clues. They are far from definitive answers as to what genetic changes lead to our advanced brain capability.

A specific set of genes that has been identified in humans is known to play a key role in the growth of the portion of our brain most involved with intelligence: our neocortex. Those genes are not present in chimpanzees, so they had to have appeared after our evolutionary split from them. To further confirm this, they are also present in another human species whose full genome is known – the Neanderthals.[26] Similarly, we have genes important for our linguistic ability and manual dexterity – two key differentiators of humans from chimpanzees – that are not present in chimps.[27] The gene responsible for jaw muscles in humans is different from the corresponding gene in chimpanzees, undoubtedly leading to the much smaller muscles in humans.[28] In chimps, these muscles are massive and their anchor on the skull causes a large depression. A smaller jaw muscle would have allowed more brain growth in the skull. These random genetic mutations, and likely many others, were naturally selected over millions of years to differentiate us from our closest living relatives.

Certainly, a larger brain would have survival advantages, assuming that it equates to higher intelligence. We will discuss intelligence in more detail in Chapter 4, but suffice it to say that brain size relative to body size does correlate with our best measures of intelligence. In humans this led to the creation of increasingly sophisticated tools, enabling better ability to hunt and prepare foods. Undoubtedly, it led to better strategies for finding and tracking game as well as securing safe dwellings and caring for young.

However, having a large brain is not all good. For one, it creates a larger head size during gestation, requiring changes in the anatomy of the female birth canal to accommodate it. Clearly, genetic mutations were selected to allow that to occur, whereas females with smaller birth canals undoubtedly had more deaths to both mother and child during childbirth. Larger brains also require more time after birth to fully mature, making childhood dependency longer. That gets back to the bipedal gait issue and other adaptations to accommodate this dependency.

A large brain also burns a lot of energy – as much as 25% of total energy consumption at rest in today's humans, even though it accounts for only 2% of body weight. That meant that caloric intake needed to become more efficient. Control of fire was one clear adaptation to accommodate this. Cooked meat and plants provide far more calories per gram and require less digestive energy than raw meat or plants. But, living meat moves around. That required us to travel over long distances on two legs to track game, overcome and kill it, and then carry it back to the other family members. Anatomically, the intestines were allowed to shrink since not as much intestinal digestion and residual waste would be required. Chimps have huge intestines and larger bellies compared to humans. Teeth, jaws and jaw muscles likewise were able to evolve. It took many random mutations between chimp-like and human to get here.

Why didn't all of this happen to chimps? Wouldn't they be better off with bigger brains? One could argue that they've done just fine living in the trees all this time. Evolution couldn't have foreseen that humans would come along and destroy their habitats millions of years later.

Evolution doesn't foresee anything. Or maybe some chimps did get the larger brain mutation but didn't also get enough of the other mutations to accommodate the bigger head and larger energy-consuming brain, so it died out in chimps. We could speculate forever about this. My view is that the answer is simple: random luck. We were randomized into the experimental leg of the first ever RCT (randomized control trial) and the chimps were the control group.

NATURE'S RCT

The RTC is the gold standard in clinical research today. With this methodology, the experimental group of patients gets a new drug or therapy and the control group is treated the old way. That's how we determine if a new treatment really works. In a sense, nature has done the same thing with brains. Through a relatively small number of random genetic changes – in both the coding genes and the epigenome – humans got larger (and presumably more intelligent) brains than chimpanzees. But we also got all of the negative consequences that go along with that, similar to the side effects or risks that come with a new experimental treatment.

The chimps were the control group, and over the past 5–7 million years both evolutionary lines were exposed to the same environments. Both groups experienced the dramatically changing environment of East Africa during that period, which included the development of the Rift Valley and multiple changes in climate that resulted in periods of drought alternating with great freshwater abundance. There was a general cooling period that led to the forests thinning out and becoming savannahs. The humans became smarter, more adaptable and more mobile, and not only dominated in Africa but migrated to dominate the entire world. They became *Homo sapiens.*

The control group stayed behind in the trees. There are no *Pan sapiens*.

OUT OF AFRICA

Getting to *Homo erectus* and the other early humans wasn't an event – it was a slow process, playing out over millions of years. There were many dead ends. Ultimately, one thread of evolution led to a larger, more intelligent brain and all of the anatomical, biochemical and physiological changes needed to support it. In turn, those led to tool development and other behavioral and cultural changes that distinguish us from the apes. We aren't quite to *Homo sapiens* yet at this point in our story, but we were getting close.

So far, everything discussed took place in Africa. Once humans could walk long distances on two legs, they walked out of Africa. These were the relatively hairless, sweaty, bipedal, larger-brained humans called *Homo erectus*, or a very closely related species. This was almost two million years ago and is sometimes referred to as 'Out of Africa 1.' But why would they leave?

Humans at that time were hunters and gatherers who wandered for a living. All hunters and gatherers have a geographic range that they cover. Over years, that range expands and changes direction in search of better food sources, or to support a growing population, or in response to competition. Sometimes changing climate, which certainly was occurring back then, pushes populations to new territories. When one considers the thousands of years it took for this first migration out of Africa, it is not difficult to cover the distances necessary to get to the Middle East and far beyond.

Ultimately, humans moved from the Middle East into Asia and later north and westward into Europe. By one million to

700,000 years ago, humans (but not *Homo sapiens*) in one form or another were pretty much distributed throughout Asia and Europe. Later, they got to the South Pacific islands and Australia. It wasn't until about 15,000 years ago, however, that the sole remaining humans – *Homo sapiens* – finally walked across a land bridge over the current Bering Strait into Alaska and then down through all the Americas or, alternatively, by boat along the coast.[29] Recent genetic and archeological evidence hints that the migration into the Americas could even have happened as early as 20,000–25,000 years ago.

Although *Homo erectus* didn't die out until as recently as 100,000 years ago, evolution and speciation of humans did not stop during this period. We know that by about 400,000 years ago, there was another large-brained human species in Europe, and later in Asia, called *Homo neanderthalensis*, or Neanderthals. The evolutionary path from *Homo erectus* to *Homo neanderthalensis* is not known for certain. Another human species found originally in Europe is *Homo heidelbergensis*, whose brain was intermediate in size between the two. This species is the first to have built artificial shelters for dwellings. There are several other candidate human species for this transition as well.

The Neanderthals were bigger than we are, bulkier, with larger bones and slightly larger brains. They were culturally advanced, lived in families, had some type of language, buried their dead and cared for their injured. Another human species that overlapped with the Neanderthals in Asia is *Homo denisova*, or the Denisovans. They were similar to Neanderthals in advancement. Genomic analyses of today's humans – as well as our current ability to extract and study DNA from a few of the fossils of Neanderthals and Denisovans – indicate that there must have been several other unknown human species on the way to *Homo sapiens*, both inside and outside of Africa.

These so-called 'ghost species' exist only in DNA sequencer printouts rather than fossils.[30]

All of the species that evolved elsewhere following the early migration of humans out of Africa emerged before the existence of *Homo sapiens*. But, they are not our ancestors.

MEANWHILE, BACK IN AFRICA...

After *Homo erectus* began migrating out of Africa, close to a couple of million years ago, groups of them did remain there and continued to evolve as well. For example, *Homo heidelbergensis* emerged in Africa, but it is not clear whether their European version evolved independently after the first migration of *Homo erectus* from Africa or represents a later migration of *Homo heidelbergensis* from Africa. Any human species in Africa could have also migrated out.

In any event, it is clear that *Homo sapiens* evolved in Africa, probably around 300,000 years ago, long after other human species left the continent and populated most of the rest of the world. Whether we evolved directly from *Homo heidelbergensis*, *Homo helmie*, *Homo antecessor* or some other related *Homo* species from that era is not fully resolved.[31] That intermediate period between *Homo erectus* and *Homo sapiens* – roughly 1 million to 300,000 years ago – is often referred to by anthropologists as 'the muddle in the middle.'[32]

Just like the emergence of the first humans, the emergence of *Homo sapiens* wasn't some singular event in time and place deserving a monument and annual celebrations. As anthropologists Sally McBrearty and Alison Brooks describe in their seminal treatise[33] on the origin of *Homo sapiens* behavior, this was no revolution. Like all evolution and the emergence of all species, the birth of *Homo sapiens* happened over thousands or even millions of generations, step by imperceptible step, mutation by mutation and small transformation by small transformation.

In less than two million years, we went from the first humans to the first *Homo sapiens* in Africa. That seems like a short time to have produced a fairly large number of human species – perhaps ten or more. It is almost certainly the case that a very small number of each pre-human and human species populated the large continent of Africa and later the even larger areas of Eurasia. Humans in total numbered in the thousands or low millions at most in those times. How these various human groups emerged as separate species is probably related to their isolation in various areas for long periods of time. That is, they all had pre-zygotic reproductive isolation barriers as a result of physical isolation.

What exactly is the difference between *Homo sapiens* and our immediate predecessor, whomever that was? Well, there is no 'exactly,' which is a problem we will deal with again in discussing *Homo nouveau*. Although there are some anatomic differences between *Homo sapiens* and other humans, most of our differences reflect brain functions in some way.

The most obvious differences were in our tools. *Homo sapiens* had the best weapons for hunting game. We made fishhooks out of bone and twine. We had the most advanced hand axes and blades. We used grindstones and digging tools. We wore fitted clothing, an obvious competitive advantage in cold weather. We created pigments and used them for cave art and decorating ornaments that were worn. Finally, the most important difference was our language. Other humans had some kind of language, but *Homo sapiens*' language was likely the most advanced and most abstract. All of these *Homo sapiens* capabilities developed over hundreds of thousands of years.

HOMO SAPIENS DOMINATE THE ENTIRE WORLD

[AT LEAST IN OUR MINDS]

However it happened, *Homo sapiens* became a reality in Africa about 300,000 or so years ago. The oldest *Homo sapiens* fossil was found recently in Morocco, whereas the previous oldest ones were from Ethiopia, dating to about 200,000 years ago. Does this mean that our speciation occurred in more than one place or simply that we were pretty mobile within Africa? There are slight variations in the *Homo sapiens* fossils found in Africa at different times and places in the period of 150,000–315,000 years ago. Humans were separated into small pockets across a vast continent. However, they moved around and probably interbred, so one could argue (and many have) that the birth of *Homo sapiens* was a multi-regional affair in Africa, with no exact time and place.

We do know we were mobile. Genomic analyses indicate that virtually all of today's humans can be traced back to *Homo sapiens* that left Africa between 120,000 and 50,000 years ago. At one time, that migration was referred to as 'Out of Africa 2.' Remember that 'Out of Africa 1,' by *Homo erectus* or some other archaic humans, almost two million years ago, seeded the earlier humans throughout the Middle East and Eurasia. We now know that there were multiple 'Out of Africa' migrations by *Homo sapiens* before 120,000 years ago. Those earlier migrations to the Middle East, Greece and other non-African sites have left no genetic trace on today's humans, and those lines probably died out for one reason or another. That last wave ultimately succeeded and, somehow, led to the extinction of all other human species. How did that happen?

Picture the scene about 50,000 years ago. Humans were in the Middle East and throughout most of Eurasia. They were Neanderthals, Denisovans, maybe some left over *Homo heidelbergensis* and possibly other human species we can only imagine from ghost tracings in genome sequences. There were no living *Homo sapiens* outside of Africa. Then, the latest group of *Homo sapiens* migrated out of Africa and, over thousands of years, spread into the territories of these older humans. Ten thousand or so years later almost all of the other humans became extinct. For example, there are no Neanderthal fossils more recently than about 39,000 years ago. What happened?

For one, we know there was contact between the various human species outside of Africa, including the *Homo sapiens*. Europeans today have about 1–2% of their genome derived from Neanderthal DNA, and in some areas of the world that percentage is higher. This is the result of at least some minimal interbreeding between the Neanderthals of Europe and the Cro-Magnons, who were the early *Homo sapiens* in Europe. The same is true for Denisovan DNA in today's humans – it constitutes as much as 3% of the genomes of people in Papua New Guinea and Australia. Similarly, there is evidence that Denisovans and Neanderthals interbred with each other to some extent. Most Africans do not have Neanderthal DNA, since the interbreeding occurred after *Homo sapiens* left Africa. The tiny amount of Neanderthal DNA found in some Africans today is likely the result of later migrations from Europe back into Africa. The exact component of Neanderthal DNA in each of today's humans differs from one person to another. In total, probably about 20% of the total Neanderthal genome is preserved in today's humans collectively.

Interestingly, the Neanderthal genome contains between 3% and 6% *Homo sapiens* DNA. Although it is likely that most of it is the result of the same interbreeding that occurred with the 'Out of Africa 2' migration, advanced genomic analysis indicates that some of it probably came from interbreeding with those earlier *Homo sapiens* migrations that died out and otherwise left no trace. In fact, it is quite possible that they did leave a trace, and perhaps some of the DNA that we subsequently got from the Neanderthals could trace back to those earlier *Homo sapiens*.[34,35] More recent statistical genomic simulations indicate that both Neanderthals and Denisovans may have interbred with even earlier human species, such as *Homo erectus*, as long as 600,000 years ago.[36] What's certain is that the more we study ancient genomics, the more complex the human story becomes.

Why we survived and the Neanderthals didn't is an unsolved mystery. Some have speculated that there was 'ethnic cleansing' back then, with the more modern *Homo sapiens* being repelled by the bulkier, less advanced Neanderthals. But there is little evidence to support this. One could also have speculated that the bigger, stronger, larger-brained Neanderthals ethnic cleansed us, but we know that didn't happen either. Also, the Neanderthals were there first. They should have been better adapted to the local climates, foods, diseases and other environmental factors that threaten species. So, something else happened.

We do know that *Homo sapiens* had better weapons. These included atlatls (devices used to throw spears), and later bows and arrows. Neanderthals didn't. Being able to kill at a distance is far safer and more effective than requiring close combat. There is no compelling evidence that these weapons were used directly against Neanderthals. Although some Neanderthal fossils show evidence of external injury, there is no way to determine if that was caused by other Neanderthals, *Homo sapiens* or other means. More likely, these weapons just gave *Homo sapiens* an advantage in hunting animals like the mammoths, which were a major food source for both species. Thus, *Homo sapiens* may have controlled the market in big game, which could have had a major negative force on Neanderthal survival.

Was our language more advanced? If so, that could have given us a subtle but definite advantage in being able to strategize as a group and communicate those strategies both short term (e.g., for hunting) and long term. That is a possibility, but the evidence for it is scant. We share the same mutations in the language-related FOXP2 gene that differentiates us from chimpanzees, but there is a different mutation in *Homo sapiens* that is not in Neanderthals. What effect that may have had is unknown, and much is still unknown about the genes for language. Computer modeling of Neanderthal brains from fossil skulls indicates that although their brains were slightly bigger than *Homo sapiens*' brains overall, their cerebellums were smaller. The cerebellum is a part of the brain that is important in language comprehension and production. Whether this had any impact on their ultimate survival is unknown.

Pat Shipman, an anthropology professor at Pennsylvania State University, writes extensively about the extinction of the Neanderthals in her book, *The Invaders*. Shipman points out that there was a major cooling climate change during that period that had weakened the Neanderthals as a species by reducing their food supply. *Homo sapiens* then invaded their territory and, like all invasive species do, put further stress on them. This could have been the final blow.

She also points out several competitive advantages the *Homo sapiens* had over the Neanderthals. One was the superior weapons mentioned already. The second was that *Homo sapiens* knew how to make bone needles and thread, which the Neanderthals did not. This enabled *Homo sapiens* to make fitted clothing rather than just draping themselves in animal hides.[37] This had to have made them more competitive during the cold winter months. Finally, and really the *coup de grace* for Shipman, was that only the *Homo sapiens* domesticated wolves. Wolves became important collaborators in both tracking and cornering large game like mammoths, making them easier to kill with their atlatls and spears. The wolves could also guard the kill from other predators.

Another theory is that *Homo sapiens* brought with them new diseases to which the Neanderthals had little immunity. That would have been true also for Neanderthals bringing new exposures to *Homo sapiens*. But since the *Homo sapiens* were coming from tropical areas containing a wider variety of pathogens than the more temperate climate of the Neanderthals, perhaps on balance we were the bigger threat. Again, this is a plausible theory with little evidence to support it. A recent study based on computer modeling suggests that a very slight decrease in fertility could easily have accounted for an extinction of the Neanderthals over a 10,000-year period, given that the starting total population size was only around 70,000 individuals.[38] Such a fertility decline could have been caused by climate change, decreasing nutrition and many other stress factors, including those induced by the invading *Homo sapiens* species.

Finally, a recent reconstruction of the Neanderthal skull shows that the eustachian tube, which drains middle ear fluid into the throat, was inadequate compared to *Homo sapiens*. Speculation is that this simple anatomical difference could have led to increased bacterial ear and brain infections, contributing to their demise.

It was probably a combination of many factors, some of which would have impacted the *Homo sapiens* as well as the Neanderthals. One key difference between the two species, however, is that the *Homo sapiens* had a continual supply line of new immigrants from Africa to replace those that may have been dying out. Perhaps, in the end, that was the deciding factor – we simply replaced them as they died out from climate change and other factors. The most recent and complete computer models clearly indicate that the intrusion of *Homo sapiens* – by whatever direct and indirect effects they had – was the cause of the Neanderthal demise.[39]

LESSONS LEARNED

The *Third Great Transformation* – the development of the *Homo sapiens* brain – happened over millions of years. It clearly accelerated about 2 million years ago, when our ancestor's brains began to grow dramatically in size (and presumably function) compared to the (other) apes. That size growth seems to have peaked about 400,000 years ago with the Neanderthal brain. Nonetheless, the *Homo sapiens* brain, although a bit smaller than the Neanderthal's, continued to explode in capability. Clearly it enabled us to somehow outsmart or at least outlive the Neanderthals and all other humans. It was the *Third Great Transformation*.

What is it about our brain that makes it so powerful? Certainly, size makes a difference. But, we don't have the biggest brains; Neanderthal, elephant and whale brains were/are bigger. We also don't have the biggest brain relative to body size; marmoset brains are bigger in that regard, for example. Neuron density (number of neurons packed into a specific volume) seems to be a factor. Our brains are extremely neuron-dense compared to most other species. It turns out that certain birds also have a high-neuron-density brain, which partially makes up for their tiny brain volume. That's why crows can use tools, recognize and remember faces and do other smart things.

However, it takes more than hardware to compute. It also takes software. There is clearly something about the organization, connections and functionality of the *Homo sapiens* brain that makes us superior. What is the algorithm we use? What is it about our software that makes our brains so powerful? It appears to be somehow related to our ability to think symbolically, using geometric and abstract figures to imagine concepts and later to express them in language. We don't believe any other humans did that. Tattersall speculates that this form of cognition is so efficient that it may account for the actual shrinking of the *Homo sapiens* brain since the time of the Neanderthals.[40] We will come back to this in Chapter 4.

As noted earlier, genome analysis has shown that our genes overlap with Neanderthals by 99.7%. That is an amazingly small difference, and is why they are our closest relatives. Somehow, that miniscule 0.3% difference led to observable (phenotypic) differences in our brains and, most importantly, in our cultures and adaptability to the environment. This is similar to *the butterfly effect* in chaos theory, in which a small perturbation in a complex system can be greatly amplified to cause major changes at a later time and place. The example in weather systems is that the flapping of a butterfly's wings on one continent could lead to a tornado later, on another continent. This seems to be what is happening in our complex genome, where tiny changes have been amplified to our new, highly competitive species.

Chimpanzees and other existing apes are not in our direct ancestral line. They are not pre-humans. Nor are Neanderthals or Denisovans in our direct ancestral line. They are humans who evolved from a common ancestor species migrating earlier out of Africa, such as *Homo erectus*. Our most likely immediate predecessor was *Homo heidelbergensis* or some close relative.

We got here through Darwinian natural selection. There were many human species overlapping in time along the way. There was at least minimal interbreeding between many of them, including *Homo sapiens*.

In summary, it took billions of years for the first life forms to evolve into pre-human ape-like creatures. It took another five million years for those pre-humans to evolve through a number of human species into *Homo sapiens*. That speciation process occurred in Africa, when relatively small numbers of humans became separated and evolved over many generations into new species of humans. That process also led to other human species throughout the world, which were ultimately replaced by the *Homo sapiens* migrating out of Africa. During the last 50,000 years, *Homo sapiens* and their advanced brain capability went from thousands of individuals worldwide living in caves to billions living in cities and virtually every land-based ecological niche.

WHY ARE TODAY'S HUMANS ONE SPECIES?

That brings us back to the species definition discussion of the last chapter. We know that a relatively small number of *Homo sapiens* left Africa in the most recent wave and eventually spread out through the entire world. Why didn't that lead to multiple human species, as it had in the earlier migrations? We know that there are regional differences in today's *Homo sapiens* that more or less correspond to our notion of races. It is a legitimate question to ask whether, for example, today's Northern European *Homo sapiens* are a different species from today's East Asian *Homo sapiens*.

Geographically, it is clear that *Homo sapiens* first entered the Middle East, where they may have encountered Neanderthals and done some interbreeding. From there, *Homo sapiens* continued to migrate throughout Asia, west to Europe and eventually into the South Pacific, Australia and the Americas. This spreading out of *Homo sapiens* covered essentially all habitable land masses on Earth, even though the populations were small. Keep in mind that for most of the first two million years of human existence, the total human population never reached 1 million individuals. It is only in the last century that our population has exploded to above 7 billion. Those early, small population groups spread across the globe and were generally isolated from each other.

It is easy to imagine speciation occurring because of this geographic isolation – that is, a pre-zygotic reproductive isolation barrier could have developed. Further, those genetically diverging populations could correspond to today's notion of races.

The first problem is that it is difficult to define what is meant by 'today's notion of races.' Are there three races – Caucasoid, Negroid and Mongoloid – as once thought? Or are there seven races – West Eurasians, Africans, East Asians, South Asians, Native Americans, Oceanians and Australians – as per one suggested classification? Whatever your definition, have such separate geographically-based populations remained stable over the tens of thousands of years since those initial migrations? That is, can we identify groups of *Homo sapiens* today that are evolving independently of all other *Homo sapiens*? If so, one would expect clear genetic separation of these populations that could reach some definition of separate species.

In fact, that is not the case. There was never stability in those separate populations. Thanks to the work of geneticist David Reich[41] at Harvard and others looking at ancient DNA from human bones from all over the world, we now have a completely new picture of what really happened after the initial migrations out of Africa. For the most part, the people who occupy virtually all parts of the world today are not the same people who originally settled there. There have been continual waves of migration in virtually every direction, including back into Africa. In some cases, this led to the replacement of original populations, and in all cases it led to a mixing of the various migrants and their genetic pools. As Reich states in his book, "We now know that nearly every group living today is the product of repeated population mixtures that have occurred over thousands and tens of thousands of years. Mixing is in human nature, and no one population is – or could be – 'pure.'"

Thus, there are no real or pure Spaniards, Brits or Indians, just as there are no races.

For example, the hunter-gatherers who colonized what is now Great Britain had a dark skin color. That population has since been replaced by massive migrations of people whose primary ancestry was from the steppes of Eastern Ukraine and Southern Russia. Most contemporary Europeans are not really 'European' – whatever that means. They originated from much more eastern locales. According to Reich, "The extraordinary fact that emerges from ancient DNA is that just five thousand years ago, the people who are now the primary ancestors of all extant northern Europeans had not yet arrived." Certainly, most of the people populating the Americas today are not descendants of those original Asians entering via the land bridge or by boat from Siberia. Instead, they are a mixture of people related to West Eurasians and East Asians who, in turn, are mixtures of other populations. The country that is known today as India is another excellent example. Reich goes on at length tracing the various migrations into and out of India, including ancient populations that no longer even exist as definable groups.

Reich and others have proved that races as commonly understood do not exist as clearly definable groupings. He cautions us, though, against concluding that genetic variations between different populations don't exist or aren't meaningful. For example, people who identify themselves as African generally have darker skin than people who identify as Caucasian. Reich compared the genetics of 'self-identified African-Americans' with 'self-identified European-Americans' and found statistically significant differences in genes predisposing to prostate cancer. That is useful information. There are many other medically meaningful genetic variations between different ethnicities, cultures and geographies.

But are these differences racial? Are 'self-identified African-Americans' a race? Everyone remembers the political flap over Senator Elizabeth Warren calling herself an American Indian. Her genome analysis showed that she indeed had some American Indian ancestry, but it could have been as low as 1/1024 parts. What race is she? We would likely find at least that amount of American Indian genetic traces in many Americans. So, what does 'self-identified' mean? Then there is Kamala Harris, the U.S. Vice President, who is the daughter of a father from Jamaica and a mother from India. Do we refer to her as African-American, Asian-American, Black, a 'woman of color,' or all of the above? In Reich's study of self-identified African-Americans, most of their genetics trace back to a West African origin, but a significant portion trace back to a European origin. Not only is the genetics unclear, but also our terminology.

We don't really have a way to characterize the genetic composition of racial groups because they don't exist. Today we only have 'populations' (the term Reich uses throughout his published work) of various ethnicities, cultures and geographies that are genetic mixtures of many previous populations. No matter what definition one uses to define a race, its genetic composition is as variable within that group as it would be between groups. There is no genetic formula to clearly define any one group as a race. We cannot today identify any sub-group or population of *Homo sapiens* that are evolving independently. That is what would be needed to define separate species.

The species problem will not go away before *Homo nouveau* is created. We'll just have to live with it. The criteria we'll use to determine the new species will be whether we can identify a group of humans who are evolving independently of all other humans. That will require verifying a reproductive isolation barrier to enable that. Certainly, no such barrier exists today among the billions of *Homo sapiens*.

GETTING TO *HOMO NOUVEAU*

Although speciation hasn't occurred since our most recent migration out of Africa, our brain has continued to evolve and, as a result, we have continued to make use of that brain to create amazing tools like computers, CRISPR gene editors and AI. Chapter 3 will outline how this tool development capability paralleled the *Third Great Transformation*.

Our next speciation will be dramatically altered by these tools. Evolution and natural selection will continue, of course. But the next species of humans will be created by us using our advanced tools over a period of decades rather than millions of years. Chapters 4–6 will discuss those tools in more detail. To meet the definition of a new species, *Homo nouveau* must be shown to be evolving independently of *Homo sapiens*. In the final chapters I'll describe one possible way that this can occur. That would be the *Fourth Great Transformation*.

When that happens, we will then have a second human species co-existing with *Homo sapiens*. That will not be unusual. In fact, for more than 99% of the time that *Homo sapiens* have existed, there was at least one other human species co-existing with them – and usually more than one. It is our current situation that is unusual. What will be unique about the *Fourth Great Transformation* will be the mechanism of speciation.

CHAPTER 3

TOOLS

"The complexity of the brain is simply awesome. Every structure has been precisely shaped by millions of years of evolution to do a particular thing, whatever it might be. It is not like a computer, with billions of identical transistors in regular memory arrays that are controlled by a CPU with a few different elements."

—**PAUL ALLEN**, co-founder of Microsoft, and **MARK GREAVES**, computer scientist, *MIT Technology Review*[42]

The evolution of tools parallels the evolution of the *Homo sapiens* brain. So, by looking at tools and their evolution, we get another view and description of the *Third Great Transformation*.

The *Fourth Great Transformation* will occur by the use of two of our most advanced tools: AI and genetic engineering. Each of these tools will be discussed in depth in the next chapters.

WHAT IS A TOOL?

The word tool often describes a physical object or instrument, like a hammer or telescope. It could also be a concept, like mathematics or accounting. In this chapter, a tool will refer to any invention or creation by any species, whether physical or conceptual. Furthermore, that tool must help that species to exist or adapt to its environment.

INDUSTRIES, AGES AND REVOLUTIONS

The Stone Age began about three million years ago and ended roughly 26,000 years ago. It is the earliest period of tool creation, where pre-humans first used simple techniques of knocking two rocks together to make a useful implement. These techniques are grouped into what archeologists call *Industries*, which simply refer to the collective production of stone tools. The earliest and crudest tools were made by the *Australopithecus* pre-humans in Africa.[43] These were simple, sharp flakes of stone used to butcher small animals and cut open bones to get at the nutritious marrow. These are called the Oldowan Industry, after the Olduvai Gorge in Tanzania where they were first found. They were also used by what some consider to be the first humans, *Homo habilis*.

Next came the Acheulian Industry, named after their initial find at Saint-Acheul in France, dating to between 1.5 million and 250,000 years ago. *Homo ergaster* and later humans demonstrated a more advanced technique in shaping and sharpening the tool into a larger, symmetrical, bi-faced hand axe. These were used for butchering, digging, woodworking and other efforts, and are found throughout Africa and Eurasia.

The Levallois Industry, first discovered near Paris, France, encompassed shaped stones that were used as weapon tips for spears. These were used from about 300,000 years ago until about 40,000 years ago, shortly after the last wave of *Homo sapiens* arrived in Europe. A version used by the Neanderthals is called the Mousterian Industry. These tools demonstrate another leap in tool creation called hafting. This is the attachment of a stone or bone tool to a wooden handle or spear, using either a plant sinew as twine or a resin as glue. The use of twine or string dates as far back as 120,000 years ago, and was later also used to make necklaces, sew clothing, and make fishing nets and bows.[44]

Finally, the most sophisticated tools of the Stone Age are called the Aurignacian Industry, which were used by the early *Homo sapiens* until about 26,000 years ago. These are thin blades made of stone, antlers or bones that were lighter, sharper and more efficient than previous stone tools used by Neanderthals and other earlier humans.

The progression of these Industries reflects increasingly complex cognitive processes in learning, planning, stone selection and preparation, and manually executing precise tasks. These initial tools required first imagining how a tool could be used and what it might look like. Then, the right kind of rocks had to be found and sometimes transported long distances. Finally, the correct techniques had to be developed to fashion the tool. By the end of this period, the stone tips were attached to spears or wooden ax handles using twine and adhesive materials. The controlled use of fire was required to produce these adhesive resins, and the spears needed to be specially grooved to accept the tips.

Significant planning, foresight and production steps were required to create these tools. Over this same period, the human brain grew from chimp size to its maximum, which was about three times larger in the Neanderthals. Clearly, the capabilities of the brain grew along with its size, as illustrated by the complexity of these tools. Improvements in manual dexterity, another manifestation of an evolving brain, were required to implement the increasingly complex, precise and symmetrical organization of the tools. Finally, language had to be evolved to communicate the tool-making knowledge to others, such as where to find the best stones and materials, how to blend the ingredients for the adhesives, and how to teach skills to others. Figure 4 shows the timeline of these Stone Age tools, along with the timelines of the species that used them.

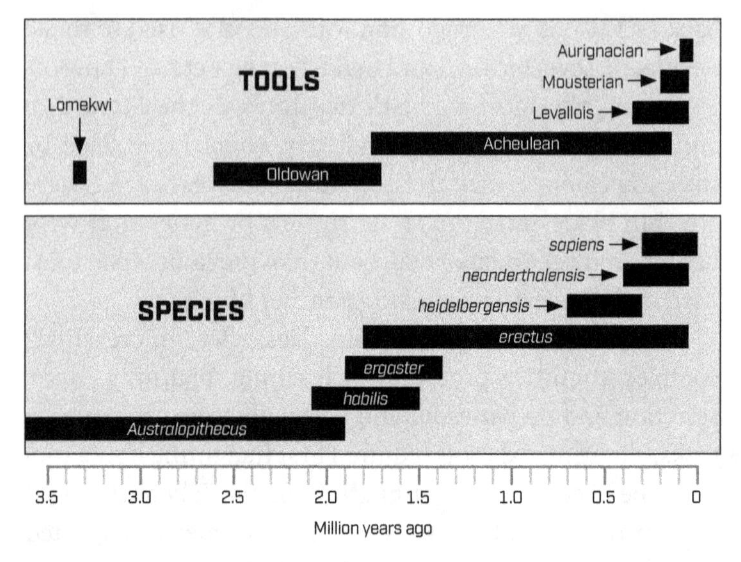

FIGURE 4 – **STONE AGE TOOLS AND HOMININ SPECIES**

GENES VS. MEMES

The Stone Age began with the immediate predecessors to humans, the Australopithecines, and ended after *Homo sapiens* were well established in many parts of the world. That's a lot of Darwinian evolution. So, is tool-making genetic? Can we find the gene for making spear tips? Were specific mutations required to move from the Oldowan Industry to the Acheulean Industry, and so forth? Yes and no.

Brain growth and processing capability were clearly driven by gene mutations, and then selected to allow better adaptability to the environment. Advancing brain capabilities were translated into specific activities, such as tool making, inhabitation of dwellings, use of fire and other cultural activities. These varied across human species and even within species in different locations. However, there is no specific gene for any specific activity.

Genes determine the capability of the brain, which is often referred to as the brain algorithm. That algorithm is the organization of the brain's neurons and chemicals that allows it to perform complex and intelligent functions – something that is not completely understood today. Memes are cultural concepts or social ideas that are transmitted from person to person. So, memes determine how the results of that brain capability are disseminated throughout a population. This transmission can be in the form of language, gestures or symbols, and they could be iconic images, songs, dances, phrases or words. Memes are communicated through visual, verbal and other sensory forms of transmission.

Unlike genes, which can only be transmitted in humans from parent to child through DNA,[45] memes can be transmitted between any two humans or groups of humans. In that sense, they are more powerful than genes. They can spread quickly and broadly through a population. Today, that is particularly evident in social media, via the internet. Memes are much more efficient than genes in passing on skills, since one person can summarize a vast amount of knowledge and pass the essentials to another person. Genes don't have any such filters or summarizers. Memes are the only essential resource that drive economies and are not consumed or diminished by use.[46] They can be reproduced endlessly, without cost.

Like genes, however, memes evolve or change in the process of transmission. There is a form of natural selection that tends to enhance and expand the influence of some memes while others fade. It can be considered a form of cultural natural selection. Although it's debatable, only humans use memes. Certainly, the techniques for creating tools and how to use them are conveyed through memes rather than genes. In fact, the concept of a tool is itself a meme.

Culture is also spread through a population via memes. As a population migrates, their culture migrates with them. This includes memes, tools, language, religion, mores, dwellings, use of fire, family structures, handling of the sick and dead, and other behaviors. So, migration is one way that tools moved over distances during the Stone Age and later periods.

There is evidence that tools were also moved by memes. Pottery was one tool that was invented shortly after the Stone Age and had many uses, such as serving as water containers. The work of David Reich, cited earlier, is instructive.[47] One form of pottery, called Bell Beaker pottery, spread rapidly throughout Western Europe about 5,000 years ago. Human migrations were one reason that pottery appeared, for example, in Great Britain. However, the movement of this pottery from Spain to Central Europe was not accompanied by population migration, but was spread by memes across bordering populations and trade routes.

Memes have taken the lead over genes in helping humans adapt to their environment, and they now surpass Darwinian natural selection in that regard. Of course, natural selection still occurs, and the *Third Great Transformation* – the *Homo sapiens* brain – is the result of it. But, that same brain has enabled us to create tools for adaptation that are now rapidly spread by memes.

THE AGE OF AGES

Once we get beyond the Stone Age, the only humans left on Earth were *Homo sapiens*. It was around 6,000 years ago that these humans first noticed something unusual in the rocks around their campfires. Sometimes a strange liquid would form that hardened upon cooling into whatever shape had contained it. These first metals were probably lead and tin, since they are liquid at campfire temperatures. Previously, humans had been aware of only the naturally occurring metals – gold, silver and copper. But these were mostly used for ornamental purposes and not as tools.

It was known at that time that copper could be shaped by pounding it with rocks, which also hardened it. Most other metals, like tin and iron, do not exist in pure form naturally, but rather as compounds contained in ores. The metal in ores is released only after the ore is heated by a process called smelting. It wasn't until the discovery of tin, however, that humans first began to smelt copper and tin together. This formed an easily molded but very strong metal alloy called bronze. What followed was the birth of metallurgy, and the Bronze Age, around 3000 B.C. For the next several thousand years, stone tools were gradually phased out and were replaced by a greater variety of bronze tools: weapons, utensils, platters, farming and gardening equipment, you name it.

Unlike the Stone Age, the Bronze Age was not accompanied by an enlargement of the human brain. Nonetheless, the advances in human culture and tool making accelerated during this period. Migrations were occurring everywhere in the world, sometimes by boat or over land, facilitated by domesticated horses and horse-drawn vehicles. This allowed the development of trade among distant populations. The wheel had come into use first for pottery creation and then for transportation and farming. Agriculture spread from the Levant (roughly the Middle East today) and Asia throughout the rest of the world. This ultimately led to the displacement of many hunter and gatherer populations and the development of cities, kingdoms, and then empires. The human brain was certainly changing during these times, but not in any ways visible in fossils.

The Iron Age is the third of the so-called Ages of mankind. It overlapped with the Bronze Age and began about

4,000 years ago, when the smelting of iron ore began to replace bronze. Iron by itself is not much more useful than bronze and requires specialized ovens to generate a much higher heat to smelt the ores. Over time, however, experimentation with carbon alloys led to the production of steel, which is stronger than bronze. Eventually, steel could be mass-produced because of the greater availability of the necessary raw materials.

By convention, the Iron Age is said to have ended about 2,500 years ago, with the proliferation of one of our most powerful tools: writing. The written record dates back to the early Bronze Age, but it didn't become widespread until the end of the Iron Age.

Just as the Industries were imprecise in timing and definition, so too are these three Ages. Clearly, the Iron Age has never ended, as evidenced by our massive steel production today.

REVOLUTIONS

Although the concept of the three Ages (Stone, Bronze and Iron) is a common convention, it relates to only the earliest phases of tool creation. The history of our so-called *cultural* revolutions overlaps with these Ages and continues to the present day.

Yuval Noah Harari, the well-known historian, philosopher and futurist, has described four cultural revolutions in the history of *Homo sapiens*: the Cognitive Revolution, the Agricultural Revolution, the Scientific Revolution and the Industrial Revolution.[48] Each of these revolutions changed our cultures and our tools.

Harari's Cognitive Revolution began about 70,000 years ago and lasted through the Stone Age, until about 12,000 years ago. This period encompassed dramatic changes in the brain's cognitive capacity as *Homo sapiens* spread throughout the world and replaced all other human species. Note that by 70,000 years ago, the human brain had already gotten as big as it was ever going to get.

But something about the brain algorithm continued to change in a dramatic way. These changes included advanced language, fictive and symbolic thinking, and complex planning and cooperation among larger groups. This drove advances in butchering tools, weapons, bone needles for clothing, boats and primitive dwellings. Other brain manifestations also became evident during this period, including cave art, body ornaments and musical instruments.

About 12,000 years ago, small groups of humans began farming and domesticating animals. This happened independently in the Middle East, China and Central America. Thus began the Agricultural Revolution, and the use of new tools like scythes, hoes, shovels, plows and animal harnesses. In fact, animals themselves then became tools, in addition to being domesticated for food. Stone and wood tools gave way to bronze and steel. Better axes, chisels and other implements were created to build animal pens, fences, granaries and dwellings for the larger settlements. Clay was baked into bricks for building stronger structures for living, praying, meeting and ruling. The wheel was used for pottery and later for carts and chariots. Pulleys hoisted water. It was a period when humans discovered and exploited the basic machines of physics.

Was our brain changing during the Agricultural Revolution? Clearly the tools were much more complex. There were also other cultural changes that imply new patterns of thought. The brain algorithm was called upon to do more because farming required planning ahead. Seasonal variations and yearly weather patterns called for optimization of seed selection, planting schedules and planning for drought or other bad weather periods. Grain needed to be stored for future low yield seasons. Water management was required. Animal care and breeding was needed to optimize meat, milk and egg production.

People now had roles and dependencies, specialization of jobs, and hierarchies of planning and management. Food had to be distributed, since not everyone directly produced it. Populations grew dramatically and clustered into larger villages, settlements, towns and cities. This required rules, planning and organization. By 4,000 years ago, there were empires and kingdoms in which most people were farmers, many were soldiers, others were craftsmen and administrators, and a few elites ruled them all. In order to collect taxes and help manage all this, one of the greatest tools of all was developed: writing.

Written history began in roughly the same place as the Agricultural Revolution – in the Levant. It corresponded with the end of the Iron Age. The Sumerians, who lived in ancient Mesopotamia in what is now Iraq, developed the first written tools in about 3500 B.C. Their writing was a system of hundreds of symbols representing spoken words or concepts, inscribed mostly on clay. Shortly thereafter, the Babylonians introduced their symbolic cuneiform writing on clay. This was followed by the Egyptians' hieroglyphic writing, using ink on papyrus. Eventually, the Phoenicians introduced a 22-character alphabet, succeeded by the Greek 24-character alphabet, which in later forms is used for most writing today.

Initially, the Sumerian writing was related to the increasing complexity of the agricultural society. Trade developed, taxes were levied, accounting became more formal, property ownership needed to be documented, and contracts were made for services. Without writing, none of this could have been managed. Soon thereafter, the long verbal history of poems, stories and memes was also recorded, and literature was born.

Sumerian writing also contained mathematical notations for arithmetic, algebra, geometry and advanced equations. Astronomy inscriptions demonstrated the earliest forms of scientific observation and methodology. The Agricultural Revolution clearly showed rapidly advancing human brain capability.

It also expanded another manifestation of human fictive thinking: religion. The great world religions began flourishing during the Agricultural Revolution. In the *Homo sapiens* brain, a distinct hierarchy had developed with God at the top, humans next and then all the other animals. This justified in human minds the inhumane treatment and slaughter of cows, sheep and chickens for food, as well as animal sacrifices to God.

Would an MRI[49] of human brains then have looked different from an MRI of the equally large pre-agricultural brains? There is no way to know. On the one hand, the changes in human cognition seem too great to have occurred without genetic changes affecting the brain algorithm. On the other hand, these changes occurred over such a short time period that natural selection wouldn't have had enough time to be the cause.

Just as the Cognitive Revolution never really ended, so too the changes brought by the Agricultural Revolution persist today. Writing, literature, mathematics, science, advanced tools, the wheel, buildings, farming, religion and countless other manifestations of the *Third Great Transformation* emerged during this period.

But the consequences have not all been positive. The varied diet of the omnivorous hunter-gatherers narrowed to a smaller array of grains and animal products. This less healthy diet made humans vulnerable to nutritional deficiencies and crop failures from drought, pestilence and other factors. Populations expanded while living areas clustered into smaller housing units. This enabled rat infestations and the easy spread of communicable diseases. Epidemics flourished. Most importantly, these cultural changes created great wealth and social inequality for the first time in human evolution. Most of the population became serfs and slaves, whom the elite ruled from mansions and palaces. Perhaps that was the worst legacy from the Agricultural Revolution. Or, was it the religious wars that followed?

In summary, the advent of agriculture brought enormous changes and all kinds of new tools: farming implements, animals as tools, writing, literature, housing, organizations, specializations, weapons, religions, mathematics, transportation, accounting, political hierarchies and much more. These were all spread via memes. The historian and anthropologist Jared Diamond, in his classic book, *Guns, Germs, and Steel – The Fates of Human Societies*, points out that the spread of agriculture itself led to the proliferation and spread of these tools. The spread of agriculture was driven by the availability of domesticable plants and animals that nature had already provided in each region. That is, it was the *presence* of these plants and animals that enabled the proliferation of agriculture, rather than some other conscious choices of humans. Furthermore, most of this proliferation went in east-west directions rather than north-south, because climate conditions allow such plants and animals to thrive at similar latitudes.

MONEY IS A SPECIAL TOOL

Throughout the Agricultural Revolution, trade expanded not only across regions but also within the ever-growing villages and towns. Workers specialized into vocations like craftsmen, administrators, accountants, service providers and other skilled professions. This required some new means to exchange goods and services. The barter system, which had prevailed for millennia, began to break down. How many barrels of wheat was a cow worth? How many olives get you a knife? What kind of olives? What if the knife-maker wanted new shoes? What if the shoemaker wanted meat from the butcher? There were too many possible exchanges and too many things to exchange to standardize bartering.

Even prior to the Agricultural Revolution, there was evidence that some type of accounting was used to indicate what was owed and to whom. This took the form of notches on a bone or knots on a cord. This form of pre-money could be considered the ancient precursor to our current electronic banking system.

In about 3000 B.C., our innovative Sumerians may have been the first to use a universal exchange medium – barley. A specific volume of barley grains called a *sila* became a unit of payment for both goods and services. This is an example of what is called *commodity money*, because the exchange medium itself has value. However, barley was unwieldy to carry around. About this time, the Babylonians defined the *shekel* as a unit of weight of barley so people could calculate in shekels directly. Later, the shekel became a coin.

Elsewhere during the Agricultural Revolution, cowry shells (shells of a sea snail) were used to exchange goods and services. These shells were a convenient tool, since they could be used for purchase, stored for future use, and could travel with you, albeit not easily. The cowry shell itself was valueless unless other *Homo sapiens* gave it fictive value. Only an imaginative brain could have pulled this off. Like mathematics, money is a fictive tool.

A couple of millennia later, metal coins came into use as money. Finally, gold and silver made their way into the tool story. Compared to bronze and steel, these soft metals make poor tools and had been relegated to ornamental roles. However, as money they soared to the top of human imagination.

Gold and silver coins stamped with the imprimatur of an emperor or king became the standard of exchangeable value. They surely beat snail shells and seed grains for convenience. Gold and silver coins again required fictive trust to have value. The coins solved the problem of portability, although there was a limit to how many coins could be easily transported. Many centuries later, paper money – called *fiat* money – solved the problem of physical portability, up to a point. Today's electronic bank transactions and credit cards are the true universal portability solutions.

Why does gold endure? Is it just tradition? Gold has useful properties to serve as money despite its relative softness. It doesn't rust or tarnish. It doesn't dissolve in water. It doesn't burn. It can be buried for centuries and would be the same when dug up, because bugs don't eat it. Gold is rare and is beautiful to the human eye. What other substance does all of that?

Even though we created money, bartering did not disappear. The Dutch purchased Manhattan Island from the local American Indians in the 17th century for trinkets. And what about house swapping today? Nonetheless, money is one of our more clever and useful tools.

WAS THERE A SCIENTIFIC REVOLUTION?

According to Harari, the third cultural revolution – the Scientific Revolution – began about 500 years ago. Note that there have been more technological advances in these last 500 years than in the previous 300,000 years of *Homo sapiens* existence. Harari's hypothesis is that this acceleration was driven by a change in attitude rather than technology – in short, due to a sudden belief in ignorance. That is, humans no longer accepted the notion that all knowledge was already available in religious teachings. And they abandoned the assumption that if something was unknown, that it was unknowable to mere mortals, or not important to know. The attitude change rejected those notions and recognized that there was much more to learn if we searched for it. Curiosity and the scientific method would uncover new truths, rather than relying on priests and sages to interpret ancient scripture. The scientific method required hypothesis testing through observation, experimentation and the application of mathematics. Harari cites Newton's laws of motion as prime examples.

I have a different view of the accelerating advances in science, technology and tools. The improvements in our brain algorithm were powered by two relatively constant phenomena:

1. The random mutation rate of DNA
2. The magnifying power of memes

Changes in our genome and epigenome occur at a fairly constant rate, with some leading to periodic, incremental improvements in brain capacity driven by natural selection. Any change in cognitive behavior is then magnified by the rapid horizontal communication of memes throughout societies. These two processes alone are sufficient to explain the last 500 years of progress without invoking some sudden qualitative change in our brain algorithm.

The rate and complexity of tool development have been increasing exponentially for millions of years. We have ample evidence of human curiosity and components of the scientific method long prior to 500 years ago. The ancient Sumerians wrote about astronomy and mathematics more than 5,000 years ago. Stonehenge, believed to have astronomical significance, dates back to that time. The Greek mathematician Euclid and the Pythagoreans demonstrated mathematical prowess 2,500 years ago. Archimedes was a scientist and mathematician around 250 B.C. Pottery and ceramics are other ancient examples of the application of early scientific methods in that it took experimentation and observation to develop the optimal techniques.

The tools already described during the Agricultural Revolution, such as smelting and the compounding of steel, also represented experimentation and observation. So, there is no reason to suspect an ignorance gene mutation or some other sudden change occurring 500 years ago to explain our ever-increasing brain capability.

Even in the centuries just prior to Harari's Scientific Revolution – the 14th and 15th centuries – there was tremendous human curiosity about the world and how to navigate around it. The science writer Lewis Dartnell describes this Age of Discovery in his book, *Origins*. Europeans, led by the Spanish and Portuguese, spread out in sailing ships to discover the extent of Africa, the Cape of Good Hope, the southern route to the Indies, and finally headed west to the New World. That's why Brazilians speak Portuguese today and the rest of Latin America speaks Spanish (and why cities in California have names like Los Angeles, San Francisco and San Diego). Not only did the explorers discover and map new lands and sea routes, they noted the patterns of atmospheric winds and sea currents, which enabled these explorations to succeed. Meteorology thus became a science critical to global navigation and world economy.

Science is partly the experimental testing of hypotheses. It is also the result of observation – sometimes carefully planned and sometimes accidental. Scottish microbiologist Alexander Fleming discovered penicillin when penicillium mold accidentally contaminated one of his bacteria-cultured petri dishes. Observations like Fleming's prompted the pioneering French biologist Louis Pasteur to declare, "Chance favors the prepared mind."

That prepared mind is the result of the *Third Great Transformation*. It represents an advanced brain capability not present in apes. Had a chimpanzee been shown Fleming's petri dish with the penicillium mold, the image would have registered on the chimp's retina and then in its visual cortex. But that is where the process would have stopped. There is no prepared mind there, since the chimp has only undergone the first two great transformations.

The human brain has been doing some type of observation that we now call science for more than two million years – since the Oldowan Industry. There was no Scientific Revolution 500 years ago, nor any other revolution. Instead, there has been a continual, incremental and accelerating rate of improvement in human brain capability since long before the Cognitive Revolution. Certainly, it is true that there have been more advances in the past 500 years than the previous 5,000 years. Likewise, there have been more advances in the past 50 years than the previous 500. In fact, there have probably been more in the past five years than the previous 50. (More on that in the next chapters.)

THE RENAISSANCE OF TOOLS

The 16th–18th centuries were times of great human curiosity and advances in observing and studying nature. We could call this either the Scientific Revolution or simply a continuing acceleration in human creativity. During these previous two centuries, European and Asian explorers had traveled the seas, discovering new lands and establishing trade routes. Their main tools were visual celestial navigation and the magnetic compass.

They also created crude maps of the world, but precise navigational maps require knowledge of the latitude and longitude of each location. Around 1730, the sextant was invented almost simultaneously by Thomas Godfrey in the U.S. and Thomas Hadley in Britain. This device solved the latitude problem. However, longitude required a precise chronometer that could maintain its accuracy on the bounding main. In 1714, the British government announced a £20,000 prize for anyone who could deliver such a marine chronometer. After multiple attempts over many years, clockmaker John Harrison won the prize in 1761. Finally, accurate maps for navigation at sea were now possible.

GLASS

When sand is heated sufficiently to melt, it forms glass when it's cooled. Like certain metals, glass was probably discovered five or six thousand years ago by accident around a campfire. The first known production of glass was for beads and ornamental objects, around 3500 B.C. in Egypt and the Middle East. A couple of thousand years later glass was molded into containers. In the first centuries B.C. and A.D. the art of glassblowing emerged. By the 13th century we had flat glass sheets for windows, mirrors and stained glass in churches. The artistry and beauty of Venetian glassware soon followed.

But glass also has the unique capability of refracting (bending) light. By the end of the 14th century inventors were creating small convex and concave glass discs to correct vision. Thus, eyeglasses were born, and just in time for the arrival of the millennium's greatest tool: books. Two important developments made books possible. In 1440, Johannes Gutenberg invented the printing press, and paper was being created by new techniques from China. Finally, humans were able to mass-produce the written word.

In the following decades, the world's recorded history went from hand-written scrolls to millions of copies of bibles and other books produced quickly and cheaply using moveable metal typesets. And, with the right set of eyeglasses, anyone could now read them.

Those eyeglass lenses bring us back to Harari's Scientific Revolution. In the 17th century, glass lens modifications led to the microscope and the telescope. These two tools started us on two amazing journeys of discovery that our most prepared minds are still attempting to unite today.

Once Antonie van Leeuwenhoek, the Dutch lens maker, looked into his homemade microscope and saw 'animalcules' in a drop of water in the late 1600s, the race was on to see the unseeable. This launched the sciences of histology (study of normal tissue), pathology (study of abnormal tissue), embryology, bacteriology, parasitology and numerous other scientific disciplines. For the first time, people could actually see those bacteria and archaea. They could see the nucleus and mitochondria in the eukaryote cells. Microscopes have been steadily improved in the past few centuries, thanks largely to better glass, lighting and tissue staining techniques. However, this incredible tool just whetted our appetite.

Human curiosity demanded more once the incredible detail and complexity of eukaryote and prokaryote cells could be seen with the tool-enabled eye. What were these cells made out of? This led eventually to electron microscopes that use a beam of electrons rather than photons (light) to achieve even greater resolution. Using what is called a scanning tunneling electron microscope, invented by IBM scientists in the 1980s, we can now see down to the level of individual chemical molecules interacting with each other.

By the end of the 19th century, most of the chemical elements – and to some extent how they could combine to form larger molecules and compounds – were known. But what made the various elements different from each other? There had to be something even smaller that caused those differences. There had long been speculation that there was some basic unit from which all things were constructed. This unit had even been given the name *atom*, at different times by different people. New tools were needed to find the answer.

But it also took science, in the form of experimentation and observation. In the late 1800s, an English physicist named J.J. Thompson discovered the electron. Later came the discovery of protons by Ernest Rutherford and neutrons by James Chadwick. But neutrons, clustered in the atom nucleus, and protons themselves consisted of even smaller particles called quarks. In the early 20th century, Niels Bohr postulated an ordered model of the atom, which led to the development of quantum mechanics. Much later, further experiments smashed particles together in huge colliders and found many more subatomic particles.

Meanwhile, at the other end of the glass tool spectrum, the telescope led to a much greater understanding of the big picture – the universe. One could now look up at the sky and see that planets had moons and rings. And we could observe that these planets, including Earth, rotated around the sun – our star. Now we could see stars that were outside our own Milky Way galaxy. And there were countless more galaxies with their own stars and planets. *Homo sapiens*' place in the universe kept getting smaller and less significant during this period, while at the same time our knowledge was expanding exponentially.

THE INDUSTRIAL REVOLUTION

Harari's Scientific Revolution began about 500 years ago with the printing press, eyeglasses, microscope, telescope, marine chronometer and other tools. About 200–250 years ago another process began that many call the Industrial Revolution. This period introduced an entirely new class of tools, although it was simply another chapter in the long history of *Homo sapiens* creativity.

The Industrial Revolution began in Great Britain, the dominant global economic power at the time, and later spread to the United States and other countries. The textile industry was then one of the largest employers in Britain. Water wheels, fed by steam engine pumps, powered machine-driven spinning wheels and later mechanical looms for weaving. Early versions of the steam engine were also used to pump water out of coal mines. Thus, the British coal industry greatly expanded and coal replaced wood as the country's primary energy source. One could say that the Industrial Revolution was powered by coal, and Britain sat atop one of the largest coal reserves in the world.

It wasn't until improvements in the efficiency of the steam engine by inventors Thomas Newcomen and James Watt in the 1700s and early 1800s that its use took off. Steamboats, railroads and finally steam-driven road vehicles revolutionized transportation. The internal combustion engine followed the steam engine, powering automobiles for at least the next century. Eli Whitney's cotton gin made the American South more economically viable and fed more cotton to Britain's textile industry. Farm equipment and sawmills became automated. Electricity was harnessed in the late 1800s and powered the newly invented light bulb, commercial electromagnets and electric engines. Overall, the use of human and animal muscle was replaced by machines.

An unsung hero during this period was mechanical engineer William Sellers and his standardization of screws, nuts and bolts, which allowed for the interchangeability of parts for mass production.

The Industrial Revolution was a period of urbanization and dramatic increases in trade both within and between countries. Large factories became the norm. Labor changed from hand-made piece creation to machine-assisted production lines. Productivity and per-capita income increased, as did the living standards of the middle and upper classes. Populations and capital markets expanded. There was increased demand for luxury consumer goods.

Josiah Wedgwood of England is a classic example of an entrepreneur who took advantage of these cultural changes. After making exquisite pottery for royalty, he marketed his services to the general public through advertising, showrooms and massive trading networks. He and others built private toll roads and shipping canals to key markets for wide domestic distribution. With the protection of the dominant Royal Navy, British merchants established a world-wide shipping network for importing sugar, spices, teas, tobacco and other commodities, and for exporting English manufactured goods, particularly to their new colonies in North America.

However, it was not all sugar and spice during these times. The average standard of living increased, but wealth inequality and poverty also increased. Factory workers toiled long hours for low wages. Child labor was exploited. Factories caused air pollution and water contamination. Much of the imported raw materials – like cotton from the U.S. and sugar from the Caribbean – was produced by slave labor.

Human communication also began to change during this period. An early use of electricity was for the telegraph. For the first time, we were able to communicate over long distances at the speed of light. The telephone, radio and television followed.

Was the *Third Great Transformation* still in progress during the Industrial Revolution? Undoubtedly, yes. Clearly, the most rapid growth in brain size of the *Homo* genus occurred between about 2.5 million years ago, and perhaps 400,000 years ago, when it peaked in the Neanderthals. That was well before *Homo sapiens* emerged and long before the Industrial Revolution. Note that the *Homo sapiens* brain size hasn't changed since the time of our earliest fossils, about 300,000 years ago. But Darwinian evolution did continue.

It is not known how much of that progress was related to memes and how much was due to genome change. There were no genetic changes in the past 300,000 years that have clearly led to brain changes. Still, that doesn't mean they didn't occur. My favorite paleoanthropologist, Ian Tattersall, believes that the rapid and deep cognitive changes that occurred in the last 300,000 years could not have been due to the slow process of natural selection.[50] Memes seem to be taking over from genes as the driving force of not only our tool creation, but our intelligence growth as well. This review of tool development clearly demonstrates increasing cognitive capabilities during those times.

TOOL USE VS. TOOL CREATION

This chapter has focused on tool creation as a manifestation of the *Third Great Transformation*. What follows are two examples where the *use* of tools that we didn't create reflect the same impact.

The first is our control of fire. We didn't invent fire, but our use of it as a tool has had a profound impact on human history. Somewhere in Africa about 1.5–2 million years ago, early human species such as *Homo ergaster* and

Homo erectus began using fire for warmth, light and possibly protection. About this time, the human brain began its amazing growth in size and energy consumption. That energy consumption required a more efficient source of fuel – cooked food, which as we discussed earlier needs less energy to digest. Fire thus powered humans through the evolution of *Homo heidelbergensis*, the Neanderthals, the Denisovans and finally *Homo sapiens.*

Homo sapiens took it from there. Fire for smelting ushered in the Bronze and Iron Ages and all their new tools. Next came the melting of sand into glass, catalyzing the expansion of the scientific method through the microscope and telescope. Steam and internal combustion engines are simply controlled fires that brought us the Industrial Revolution. On the negative side, controlled explosions ignited by fire gave us bullets, cannons and artillery. Finally, we got jet engines, supersonic transport, rocket engines and travel to the moon. All from fire.

The second example revolves around the electron. Again, we didn't invent it, but we sure use it. Electricity revolutionized communication, transportation, entertainment, energy and all of those light bulbs and appliances we now take for granted. Most significantly, it is the basis for the digital computer, cell phones, the internet and, ultimately, AI, which will drive the *Fourth Great Transformation.*

THE EXPONENTIAL CURVE

Once we get into the 20th century and beyond, we are so far out on our exponential curve of tool creation that it is difficult to pick out all the important ones. Table 2 is an attempt to list the major tool categories from this period and to give some examples of each. What dominates the 20th century of tools is the emergence of the digital computer, the internet and Moore's Law. Yet, as stated at the beginning of this chapter, the human brain and a computer do not work the same way.

In 1965, semiconductor researcher Gordon Moore, the future CEO of Intel Corporation, predicted that computing power would double every two years. He believed this would be true for the next decade, but it's mostly been true for the next *five* decades and counting. There is no end in sight, with biocomputers and quantum computing on the horizon. It is the Information Industry, Revolution and Age all rolled into one. Looking back at Figure 4, it took about 3 million years to perfect the stone tool. If we were to superimpose the progress of tools since the introduction of agriculture, they would all aggregate into a tiny sliver of that graphic. Tool creation – and the human brain that drives it – is clearly on an exponential curve.

The *Third Great Transformation* – the development of the *Homo sapiens* brain – is the most impressive example of Darwinian natural selection, and tool creation is its most important manifestation. We don't need to be the fastest animal when we can drive cars and fly airplanes. We don't need to be strongest animal when we have front-end loaders to do the lifting. We don't need the best vision when we have eyeglasses, microscopes and telescopes. We needn't possess the best hearing when we have amplifiers. And, we don't even need to be the smartest (although we already are) when we have AI.

The purpose of this review of tools is simply to demonstrate the parallels between increasing brain capability and tool creation – both manifestations of the *Third Great Transformation*. The next chapters will focus on two of our most advanced tools: AI and genetic engineering. AI represents a potential merger of a tool with the brain and, coupled with genetic engineering, will lead to the *Fourth Great Transformation* and the emergence of *Homo nouveau*.

CATEGORY	EXAMPLES
Transportation	Jet Engines Global Positioning System Radar Helicopter
Electronics	Digital Computer Electronic Calculator Transistor Integrated Circuit Artificial Intelligence Compact Disc Bar Code Liquid Crystal Display Computer Mouse
Communications	Cell Phone Arpanet Internet World Wide Web Protocol Wireless Communication Satellite Communication Fiber Optics
Entertainment	Talking Motion Picture Radio Television
Energy	Atomic Power Solar Power
Weapons	Missiles Atomic Bomb Hydrogen Bomb Modern Submarine

CATEGORY	EXAMPLES
Medicine	Imaging Technologies Artificial Limbs, Joints Drug Development Genetic Engineering Radiation Therapy Dialysis Electrocardiogram Defibrillator Pacemaker
Household	Vacuum Cleaner Microwave Oven Polaroid Camera Digital Camera Teflon Pyrex Scotch Tape Photocopier Ballpoint Pen Fluorescent Light
Science	Lasers Space Travel, Space Station Electron Microscope Genomics Paleogenomics Proteomics Brain Connectome Mapping Radiocarbon Dating Particle Accelerator/Collider Quantum Computing Hubble Space Telescope Nanotechnology

TABLE 2 – **SOME TOOLS OF THE 20TH CENTURY AND BEYOND**

CHAPTER 4

AI: HUMAN INTELLIGENCE IN A COMPUTER

"No, I'm not interested in developing a powerful brain. All I'm after is just a mediocre brain, something like the president of the American Telephone and Telegraph Company."[51]

—**ALAN TURING**, the father of AI

AI will be a key tool in the development of *Homo nouveau* in the *Fourth Great Transformation*. Its role in that regard will be described in Chapter 6. But first, in this chapter, we will explore exactly what is meant by the term *artificial intelligence*.

Many people have gotten their understanding of AI from popular movies like *2001: A Space Odyssey*, *The Terminator*, *Her*, *Ex Machina* or Stephen Spielberg's 2001 film, *A.I. Artificial Intelligence*. In these movies the AI is either embodied in a humanlike robot or in a computer that communicates through voice or text. These all demonstrate a state called *artificial general intelligence*, or AGI, which is when AI is equal to human intelligence. We have not achieved this yet, so these movies are still science fiction. Sure, there will eventually be some lifelike robots, but they probably won't be like the beautiful and sexy Ava in *Ex Machina*.

WHAT IS AI?

In 1955, four computer scientists[52] popularized the term *artificial intelligence*, which they said was based on "the conjecture that every aspect of learning or any other feature of intelligence can in principle be so precisely described that a machine can be made to simulate it." This was in a proposal they made to the Rockefeller Foundation for a summer workshop to be held the next year at Dartmouth University. These illustrious, though perhaps overly optimistic, computer scientists stated this in their application:

> "An attempt will be made to find how to make machines use language, form abstractions and concepts, solve kinds of problems now reserved for humans, and improve themselves. We think that a significant advance can be made in one or more of these problems if a carefully selected group of scientists work on it together for a summer."

All of that was promised, within two months, for $13,500. How could the Rockefeller Foundation refuse? Their statement still sounds like hubris, more than 60 years later, but the four scientists are deservedly considered pioneers of AI. They conjectured that non-artificial intelligence – human intelligence[53] – could be emulated by a computer. But is AI simply human intelligence in a computer? It may not be that simple.

Dictionaries define intelligence as the ability to reason, understand and learn. So, would a good definition of AI be a computer's ability to do those things? Perhaps, but that sets a pretty low bar. It all depends on what we mean by those terms. Even computers 50-plus years ago had some form of reasoning in that they could calculate things.[54] They understood their inputs (which came in many forms), computed answers in a reasonable way, and then 'learned' things by storing and retrieving their results in memory or storage devices. However, most people would not consider any of this AI.

The phrase AI is used to describe many human cognitive functions that computers already perform, like face recognition, game playing, search engines, speech recognition, chat programs, driving a vehicle and many others. These functions could be categorized as pattern recognition, strategizing, decision-making, and rule- or algorithm-making. Which of these constitutes AI? All? Some? It depends on whom you ask.

But before trying to define AI, let's see if we can define human intelligence.

WHAT IS HUMAN INTELLIGENCE?

There are many characteristics of intelligence that differentiate humans from other animals. One is the ability to create tools, as discussed in Chapter 3. As noted, other species, like chimpanzees and some birds, can use a rock or a stick as a tool and even shape sticks into hooks. But this level of tool creation and use pales compared to building computers, splitting the atom and flying to the moon.

Another differentiator is our language. Again, many species have some type of language, but none has our extensive breadth of vocabulary and syntax. More importantly, no other species appears to have our type of fictive language, which enables thinking about and communicating imaginary concepts like religion, countries, corporations and mathematical theorems. In Chapter 3, we introduced the notion of memes. Memes can be in the form of language, signs, symbols or cultural concepts. They certainly reflect human intelligence. Other animals appear to have capabilities that could communicate memes, as described in Chapter 3, but humans have taken this to a new level.

What about consciousness? Consciousness is also difficult to define but is generally considered to be an awareness of self, our thoughts, feelings and surroundings.[55] Although debated, some other animals appear to have consciousness.

Is there something different about human consciousness that is a part of intelligence? Scientist John Hands, the author of *Cosmosapiens*, thinks so. He coined the term *reflective consciousness*, which he believes is a capability unique to humans. He defines this as "the property of an organism by which it is conscious of its own consciousness, that is, not only does it know but also it knows that it knows." We not only get angry, but we know we are.

Consciousness is a vast topic that gets into metaphysical theories of the universe, quantum mechanics and other issues too far afield from this conversation. For now, let's just say that we think about consciousness the way we do, and that in itself may indicate a unique part of our intelligence.

The psychologists Michael Tomasello and Malinda Carpenter point to studies in chimpanzees and human children that demonstrate another manifestation of intelligence.[56] Both children and chimps can follow the gaze of an adult human to try to discern what the adult is looking at. But only children use gestures and nods to communicate that information to other children. They call this *shared intentionality.* They claim this ability is unique to humans and is a major factor in our superior ability to collaborate.

Some consider creativity – the ability to conceive of new concepts, art forms, music or other memes – as the unique defining characteristic of human intelligence. Brian Greene, a brilliant theoretical physicist and mathematician, argues in his book, *Until the End of Time,* that art in all forms has played a critical role in human natural selection and has increased our fitness and survivability. He states, "Neither Einstein's relativity nor Bach's fugues are such stuff as survival is made on. Yet each is a consummate example of human capacities that were essential to our having prevailed."[57] Pablo Picasso is considered a brilliant artist. What exactly was his brilliance? I've seen computer-generated art that appears to me to be equally abstract and interesting.[58] Who is to judge and how do we judge? In fact, it was Picasso who said, “Computers are useless. They can only give answers.”[59]

Finally, there is the work of medical information scientist Marsden S. Blois, published in the July 24, 1980, issue of the *New England Journal of Medicine*, in a paper called “Clinical Judgment and Computers.” The figure below is taken from this article.

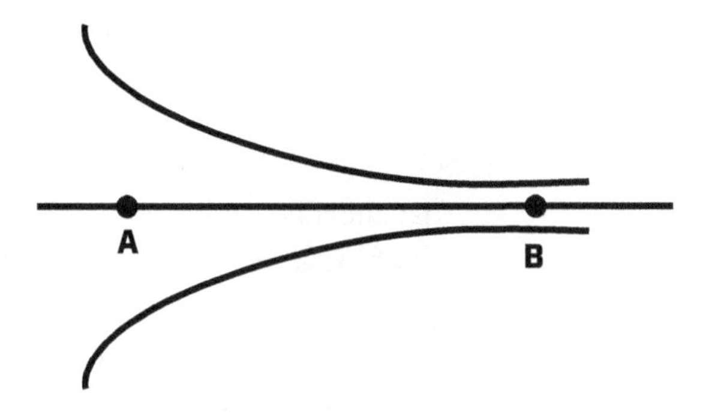

From: Blois, M. 'Clinical Judgment and Computers,' *NEJM* 303:192, 1980

FIGURE 5 – **COMMON SENSE**

The funnel shows the decreasing cognitive span of a physician during the diagnosis of a patient. Point A initially indicates that any diagnosis is possible. The physician, with his or her broad knowledge of medicine *and of the world in general*, interacts with the patient by talking (taking their history), examining, and observing their behavior. This progresses along the timeline toward Point B, narrowing down the possibilities to a reasonably small number. This is called the *differential diagnosis*. Note that getting the single correct diagnosis from Point B may still require laboratory tests and other diagnostic procedures, as well as very specific knowledge of that disease spectrum. Blois's contention was that physicians are far superior to computers at Point A, but computers – if programmed properly with the right rules – are superior to physicians at Point B. Blois published his paper in 1980, the dark ages of AI, but his observations still hold true today.

The funnel could represent any domain of knowledge, not just medicine. Somehow, the human brain is born with the capability to get to Point A just by living in the world. This is called common sense.

A related example is face recognition. Even infants recognize human faces. It has taken us decades of computer science development to create a software program that can match a five-year-old in face recognition. There's something innate going on in the human brain that we're trying to emulate in a computer, but that effort is proving very difficult.

Tool creation, fictive thinking, reflective consciousness, shared intentionality, common sense and pattern recognition are all parts of intelligence. These are not all unique to humans, but we have taken these abilities to new levels. How do these abilities relate to reasoning, understanding and learning? Defining human intelligence is quite complicated; there is no simple definition. It is no wonder we have difficulty defining AI when we can't even define what it is supposed to be emulating.[60]

GAME PLAYING PARALLELS PROGRESS IN AI

Just as tool development characterized the progression of human intelligence, computer game playing has done so for AI. In 1997, the world was shocked when IBM's Deep Blue computer beat the world champion chess player, Gary Kasparov. The event was heralded as the coming of the age of AI, when computers would be as smart as people.

Kasparov had actually beaten Deep Blue the year before, so his loss was doubly shocking to him. He took it quite personally, and in retrospect complained that he really lost the match psychologically rather than technically. He was upset that IBM would not allow him to analyze previous Deep Blue games, something that was always important in his preparation against experienced competitors. He was also unnerved when Deep Blue crashed several times during the match and had to be rebooted. You couldn't reboot a human. Finally, he considered chess to be as much a contest in psychology as skill and found it troubling to play against an opponent with no psyche.

We now know that Deep Blue wasn't really all that smart and didn't rise to the level of human intelligence. It ran a *brute force* program that looked ahead by trying out millions of combinations of future moves. It could analyze 200 million moves per second. What it demonstrated was computer speed, not intelligence.[61] Neurons are slow compared to computer chips. Humans couldn't possibly look that many moves ahead within the time constraints of a chess match. In fact, such brute force isn't how the human brain even plays chess. Although a chess player does look ahead, he or she is really seeing patterns in ways that cannot even be described – let alone programmed into a computer. One of the greatest chess masters of all time, Emanuel Lasker, described the special intuition needed for the game in one of his books, *Common Sense in Chess*.[62] No, Deep Blue was not as smart as a human. It was simply faster. It didn't have the common sense and special intuition that the best chess players have. However, there is no dispute that it was better at chess.

But here is a key fact about chess playing software. When the software is used as an *adjunct* to a human it is even more powerful. That is, a good chess player using the software to decide the next move is actually superior to the software alone. This concept is more broadly described as *intelligence augmentation* (IA) rather than AI.[63] As will be discussed in later chapters, IA will be key to enabling the creation of *Homo nouveau*.

In 2011, another milestone in computer game playing was achieved when IBM's Watson computer beat the reigning Jeopardy game show champions Brad Rutter and Ken Jennings. To most lay observers, this brilliant intellectual achievement demonstrated true human intelligence. I remember reading about a question that Watson correctly answered from the following clue: A LONG TIRESOME SPEECH DELIVERED BY A FROTHY PIE TOPPING. Watson pressed its mechanical button first and correctly answered: WHAT IS A MERINGUE HARANGUE? That is simply amazing. Not only did Watson understand natural language and have information about pie types, but it understood the subtleties of a pun and could create a clever rhyming phrase. It knew how Jeopardy works and what kinds of answers it expects. And, in response to Picasso's criticism, it knew how to ask questions, since that is the only form of response that Jeopardy allows.

Another question in the same competition, however, revealed a strange flaw. In the category 'Name that Decade,' the clue was: THE FIRST MODERN CROSSWORD PUZZLE IS PUBLISHED AND OREO COOKIES ARE INTRODUCED. One of the human competitors answered first: WHAT IS THE 1920s? That answer was incorrect, and Watson quickly responded with its own answer: WHAT IS THE 1920s? How dumb for a supposedly smart computer, after 'hearing' that that answer had just been declared wrong. Anyone or anything with common sense would know not to repeat the same mistake. That was not the only dumb mistake Watson made, but it still eventually won the contest.

What is going on here? Certainly, the software in Watson was a major advance from Deep Blue. Consider all the different cognitive-like functions the computer needed to perform in order to play Jeopardy. First, it had to understand the English language, with all of its ambiguities, synonyms and homonyms. Also, Jeopardy's clues often use nuanced phrases, humor, irony, metaphors, rhyming, puns and clever mash-ups of ideas that make it difficult even for humans to understand. Strategies are also needed for the size of bets and the categories to choose. And after 'hearing' the clue, the computer must quickly search its vast database for answers, determine which answer is most likely correct, decide if it is worth risking an answer, and, if so, press the buzzer before the human competitor does. All of these actions must happen in scant seconds.

This seemed like an impossible set of tasks, so the Watson developers did not create a single program. They created many of them, each specialized in different aspects of the problem. All the programs worked in parallel, similar to how the human brain solves problems. Each program fed pieces of the solution to the other programs. During the development process, as Watson evolved, certain flaws cropped up that required even more specialized programs. For example, at one point its vocalization software developed a bit of speech impediment, causing it to add a 'd' after an 'n.' For instance, it pronounced Pakistan as *Pakistand* and Bhutan as *Bhutand.* Not only was this embarrassing, but these were wrong answers. Watson's programmers didn't know why it was doing this, but it was easy to fix by adding another module.

All of these programs were 'taught' the idiosyncrasies of Jeopardy clues by training them from an archive of 180,000 clues from previous games. Watson's creators even studied the future opponents, Ken Jennings and Brad Rutter – how often did they hit the buzzer, what was the average time to respond, how often were they correct, how did they bet? This is analogous to how Gary Kasparov prepared for major chess matches.

The end result was a complicated set of algorithms operating on thousands of parallel processors fine-tuned over years to not only play Jeopardy, but to beat these particular opponents. It contained in its memory more than 200 million pages of information from the internet. Was this AI, or do we just chalk it up to better information access? It succeeded in its goal of emulating many human cognitive functions. But, most would say it was not human intelligence. Watson lacked common sense. It was not aware of what it was doing. It experienced no joy in winning. It made really dumb mistakes. On the other hand, humans also make dumb mistakes, so that doesn't mean you don't have human intelligence. Computers simply make different kinds of dumb mistakes. So, should computers only make humanlike mistakes?

Whether Deep Blue or Watson is really AI is debatable, and our frustration in trying to define it is perhaps best illustrated by what is called the *AI effect*, or *Tesler's Theorem*. Larry Tesler, a computer scientist and former executive at Amazon and Yahoo, is credited with defining AI as whatever a computer hasn't yet been able to do.

AI engineers continued to advance their techniques, as reflected in game-playing programs. These same techniques were used in every area in which AI was applied, so their progress in game playing reflects their progress in all fields. The next challenge was the strategic board game Go. Figure 6 is a snapshot of a game in play. Each player in turn places a tile on an intersection. Completely surrounding the opposing player's tile or tiles allows the player to remove those tiles. The object is to have surrounded the most territory by the end of the game. This ancient Chinese game, created more than 2,500 years ago, is considered more complex than chess, requiring fundamentally new approaches to AI.

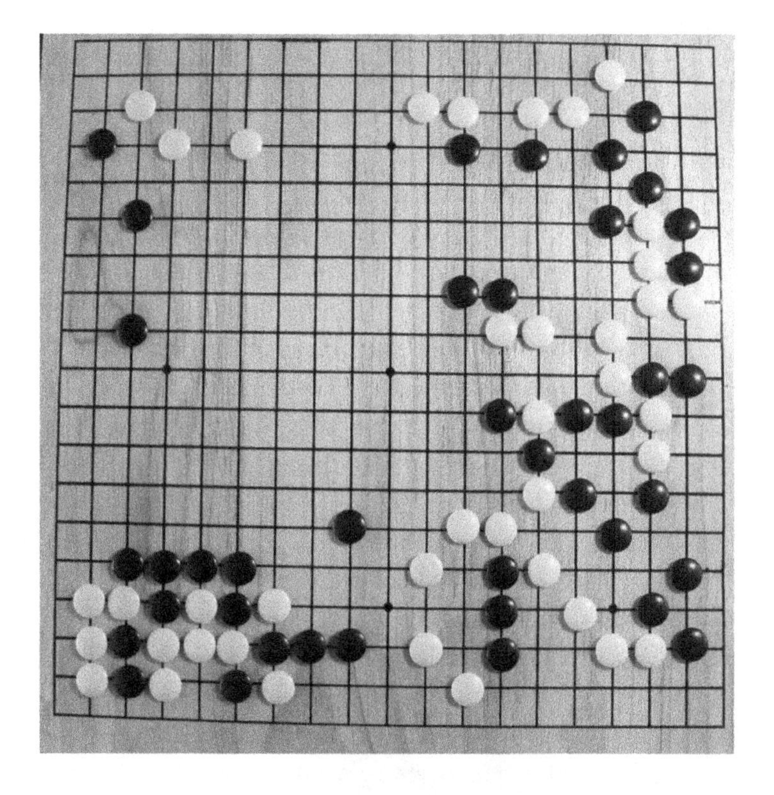

FIGURE 6 – **THE GAME OF GO**

The computer program AlphaGo was created by a company called DeepMind, founded in 2010 by Demis Hassabis, Shane Legg and Mustafa Suleyman. Google acquired the company in 2014 and, coupled with Google's technical expertise and money, DeepMind improved its AI algorithms to exceed human ability in various computer games, including Go. AlphaGo utilizes a software technology called *deep neural networks* and illustrates the next advance in AI. These complex programs try to mimic the human brain by using many computer processes operating in parallel. The human brain consists of many networks of neurons that work, autonomously and simultaneously, to perform virtually all of our cognitive functions. Most of these processes occur at a subconscious level, and only periodically reach conscious perception. These processes make decisions and initiate actions that occur even before the person is aware of them.

One of the key features of the initial AlphaGo software was its ability to learn, called *machine learning*. This is another key component of modern AI. By playing multiple games of Go against humans, it learned the best strategies over time. What is meant by 'learning' in this sense is simply that the software improves over time in its ability to accomplish its goal – whatever that is, like winning a game or correctly identifying an object in a photo – by repeated running of the software. By 2016, AlphaGo was able to beat the 18-time world Go champion, South Korea's Lee Sedol, four games out of five. (We will return to this competition later.) This was a devastating blow, not only to Sedol but to the entire South Korean population, which prides itself on Go mastery. During these matches, AlphaGo made some surprising moves that defied the logic of even the most advanced Go experts, yet they proved masterful. Gary Kasparov said after his experience with chess AI, "To be good at anything you have to know how to apply basic principles. To become great at it, you have to know when to violate those principles."[64] AlphaGo certainly violated a lot of Go principles, all the way to the winner's circle.

A later release of AlphaGo, called AlphaGo Zero, does not even need to learn by playing real games with humans. It can start from a completely blank level of knowledge of Go and start playing games with itself. By doing this, it surpassed the capability of the original software in a matter of days. An even later iteration called AlphaZero demonstrated the same capacity across multiple games, including chess. This ability to self-learn and self-program without the assistance of humans raises serious concerns for the future, as will be discussed.

Chess, Go and Jeopardy are games characterized by what is called *perfect information*. That means that nothing is hidden, and each player has access to all of the information. For example, in Go, as in Chess, each player sees the same board and witnesses the other player's moves. The next game playing hurdle for AI was to tackle games with imperfect information – that is, where some information is hidden from the opposing players. Similar situations occur in many instances other than game playing to which AI is applied, such as negotiations and encryption. Poker is such a game, since most types have one or more cards that are not visible. Texas Hold'em is one of the most widely played types of poker, and well-publicized high stakes competitions are popular on TV.

AI researchers from universities in Canada and the Czech Republic collaborated to develop a deep neural network program called DeepStack.[65] Special software techniques were created to deal with hidden information and also the betting strategies, which are a key component of poker. These include the size of bets, raising bets, holding and folding and, that most characteristically human technique, bluffing. DeepStack was trained by playing 10 million randomly generated Texas Hold'em games. In December 2016, DeepStack was tested against 33 professional Texas Hold'em players who were recruited with the offer of large cash prizes. After four weeks and 44,852 head-to-head (two-player) games, DeepStack emerged the grand winner by a huge statistical margin. The computer could out-bluff the best of them when it needed to. Would a casino ever be safe again?

But in most casinos – and especially in the big annual Texas Hold'em shoot-outs in Las Vegas – the game is played with six players simultaneously. This adds considerably greater complexity to each hand. A new group of professionals was recruited in 2019 to go up against a new poker-playing program called Pluribus.[66] The developers of Pluribus were Tuomas Sandholm, a professor of computer science at Carnegie Mellon University, and Noam Brown, a research scientist at Facebook AI. Each of the professional players in this competition had previously won at least a million dollars playing Texas Hold'em. The prize money was $50,000 for them to play more than 10,000 hands of six-person poker against Pluribus. It was five of the world's fastest poker gunslingers competing with Pluribus. Once again, though, the humans were outwitted and outbluffed. Pluribus knew when to hold 'em, when to fold 'em and when to bluff 'em.

Part of poker playing by humans is the ability to look your opponents in the eye and psych them out – figure out their patterns of play, watch for hesitations or nervousness or 'tells' (changes in mannerisms or behavior), know when they're bluffing, scare them with big bets, sucker them with 'donk' bets[67] and unleash all sorts of human tactics. Chess players exhibit some of these same characteristics. So, how do you psych out a computer? You can't. And how did Pluribus do that to the other players? It played millions of games and learned from them so well that it didn't need psychology. This is another example where AI is clearly different from human intelligence.

Pluribus is significant for a number of reasons. Its self-learning deep neural network software ran special algorithms that made it more efficient in real-time execution than Deep Blue, Watson and AlphaGo. Its training was faster and cheaper than previous programs. It reduced the number of options needed for each decision in situations with hidden information. Many of the real-world problems that AI will be called upon to solve, like cybersecurity, are of this nature.

These new efficiencies are extremely important given the magnitude of the possibilities – the 'space' – that needs to be considered when making a decision. Remember that Deep Blue looked ahead many millions of possible moves in its brute force analysis, yet that didn't cover the total possible moves in a chess match, which is estimated to be 10^{120}. That number is orders of magnitude greater than the total number of atoms in the entire universe. Go's number of possible moves is 10^{320}. The number for six-handed Texas Hold'em is probably even higher. That is why AI software cannot succeed by brute force. It must be much more intelligent – just as we know that the human brain does not use brute force when playing these games. The Pluribus developers made a breakthrough in reducing the decision space, which is highly significant and can be applied broadly.

There are other approaches being developed to assist AI in reducing the decision space. AI software learns whether any AI decision is a good one based on the ultimate goal or outcome. For instance, if a certain move early in a game usually leads to a win, then that move is evaluated highly. Over time, the software can detect that pattern and therefore 'learn' which moves are best. Sometimes, though, truly bad moves are made in games that are ultimately won. Savielly Tartakower, the great chess grandmaster of the early 20th century, said that the victor is the player who makes the next-to-last mistake. Somehow, the AI system must detect when a 'bad' move is really a good one on the road to ultimate victory.

For example, in chess a player has to evaluate each move in relation to the short-term impact on the value of the remaining pieces, as well as its long-term impact. A rook (castle) is more valuable than a bishop. So, sacrificing a bishop to win a rook would usually be considered a good move. However, there are unusual circumstances where sacrificing a rook for a bishop is the correct move. The software must evaluate huge numbers of games to distinguish between the really bad moves and the good sacrifices.

Computers can do that in minutes or hours, whereas it would take humans years to do the same thing. But even so, computers can't evaluate 10^{120} possibilities.

An interesting new approach in chess is being developed that could improve this evaluation function: allowing the computer to read commentary from experts during the games it is learning to play. Yes, you read that correctly: let the computer learn from what a chess expert has to say about each move. This sounds far-fetched, but it is beginning to happen.

All championship matches at the grandmaster's level have such expert commentary. A team in Britain has developed a chess program called SentiMATE[68] that uses AI-based natural language processing to rank expert comments about any move as positive, neutral or negative. Early testing of this software is promising, although it still couldn't beat AlphaGo by itself. It does suggest a future where computers also learn from humans while learning the games.

To finish this discussion of game-playing AI, the question of teamwork may be the final hurdle. In every example so far, the AI software played as an individual. In chess, Go and heads-up Texas Hold'em, it was a single copy of the AI program against a single opponent. In Jeopardy, Watson played two opponents. In the Texas Hold'em grand final, Pluribus played five opponents simultaneously. In all of these cases, there was only one copy of the AI software in competition. But what if the AI had to play on a team? That raises all sorts of new problems. How do you divide up the workload? Who is the leader, if anyone? How do you coordinate with the other members and strategize? When should a player sacrifice itself and 'take one for the team?'

There is a computer video game called Dota 2 that can require teams of up to five players to compete against each other. The online battle game is heavily graphic, very complicated, fast moving and quite bloody (virtually).

Each year, there is an annual competition with large financial prizes that attracts professional teams. In April 2019, the company OpenAI, founded by Tesla's Elon Musk and others, challenged the 2018 national Dota 2 champions. They used a Dota 2 algorithm they developed called Open -AI Five. This algorithm trained by playing 250 years of games against itself every day for ten months, for a total equivalent of 45,000 training years. It is estimated that a human takes as much as 20,000 hours of Dota playing to become a professional. OpenAI Five was playing about 110 human training lifetimes of Dota every day to learn. On the one hand, it is amazing how much learning a computer can do in a day. On the other hand, it is amazing how relatively little training the human brain takes to be equivalent to a computer.

In a difficult best-of-five match, pitting five copies of OpenAI Five against the human championship team of five players, OpenAI Five won the challenge and established yet another milestone in AI game playing – collaboration. Now, we're really getting into the sweet spot of human intelligence. As in every game playing example discussed so far, this software required further advances in AI techniques that will be applied in many real-world environments.[69]

HAVE WE REACHED AGI YET?

It continues to be a challenge for me to define AI.[70] All of the game-playing software programs we've discussed are, in my opinion, a form of AI. They all emulate some aspect of human thinking. AI software of one type or another already pervades our lives. It controls the search engines on the internet. It is responsible for those uncannily relevant pop-up ads on Amazon and Google. It converses with us through Siri and Alexa. It drives speech and face recognition and also drives automobiles. It performs in call centers. It buys and sells stocks. It provides security for our credit card transactions. It is everywhere. It outperforms humans in many tasks. But is it human intelligence? And if it is human intelligence, is it equal to the real thing or just some aspect of it? AGI implies that the computer equals human intelligence, not just emulates some aspect of it. Since we have been unable to precisely define either human intelligence or AI, how would we judge that?

In 1950, the brilliant Alan Turing proposed one way to determine when a computer matches human intelligence, now called the Turing Test. Common definitions describe this as a person, acting as an interrogator, who submits questions to both a computer and a person, each isolated from the interrogator. The answers are communicated back by typed text. If, after a period of time, the interrogator can't correctly identify which answers come from which, the computer has then passed the test and is considered as intelligent as a human.

However, the test described in Turing's original article was a bit different from that. His original test was what he called *the imitation game*. A man and a woman were secluded into two separate rooms, and the interrogator's goal was to determine who was who from the typed answers. The man and woman were instructed to intentionally disguise their gender identity in crafting their answers. The Turing Test then consisted of substituting a computer as one of them and seeing if it was equally capable of fooling the interrogator. Turing predicted that by the end of the 20th century (50 years later), "an average interrogator will not have more than 70 per cent chance of making the right identification after five minutes of questioning." That is, the computer would be as good as a human in fooling the interrogator.

Whatever version of the Turing Test one uses – and there have been many over the years[71] – it has been criticized as inadequate. Who would be the average interrogator? Who would be the persons being interrogated? The test is too imprecise. Which of the many definitions of human intelligence is really being tested? What questions would be asked? If you limited the questions to chess moves, IBM's Deep Blue would have passed the test long ago because it could more accurately pick the right moves, yet no one considers it to have human-level intelligence. One could easily determine which was the computer by asking it to do a complex calculation quickly – it would beat the human, hands down. So, one would need to strictly limit and structure the types of questions to make this a fair test. An excellent review article by researchers at the University of California, San Diego, entitled "Turing Test: 50 Years Later," described the many flaws of the many versions. It concluded by saying, "We believe that in about fifty years' time, someone will be writing a paper titled 'Turing Test: 100 Years Later.'"[72]

I have proposed a different test, called the Blois Test. Looking back at Figure 5, we could rate each AI program as to where it fits on the continuum between Point A and Point B. When an AI program is judged to be at Point A or to the left of it, it would be considered equal to human intelligence. Certainly, Deep Blue is at Point B. Watson is a little to the left of B, but far from Point A. In fact, all of the game-playing programs mentioned are near Point B. But there are flaws in the Blois Test as well. Who will make the judgments as to where AI falls in that continuum?

It is generally agreed that no computer today would pass any reasonable version of the Turing Test, the Blois Test or any other test that purports to determine if a computer's intelligence is equal to ours. Although we don't have a way to definitively define AGI, or to test whether it is achieved, it is safe to say that we aren't there yet.

SO, WHERE ARE WE?

No matter how we measure it, AI programs are getting smarter with each new iteration. In narrow activities like game playing, computers already perform better than humans. AI is now expanding into areas that require extraordinarily high levels of education and expertise. For example, there is now an AI program that exceeds the accuracy of radiologists in interpreting some types of X-rays. Another program has been shown to diagnose breast cancer better than pathologists in reading certain biopsy slides. Brain surgeons are now using AI to look at biopsies in the operating room to make decisions in minutes rather than sending frozen sections off to the pathologist and waiting 30 minutes for the answer. These are amazing achievements, but they certainly don't replace radiologists or pathologists. In another example, an AI program developed at the Massachusetts Institute of Technology can distinguish a COVID-19 patient from a non-COVID patient just by analyzing a recording of a person's forced cough into a cell phone. This program correctly detected 98.5% of all patients known by laboratory studies to have COVID-19, and 100% of such patients who were asymptomatic.[73] That will be useful as a screening test, but it won't replace physicians, who will still need to sort out false positives and treat and advise the patients.

So far, all AI applications in medicine are IA (intelligence augmentation) and are likely to remain so for decades. These AI tools are narrow in scope. We could also point out that a radiologist can't read pathology slides very well. Einstein couldn't either. Do we call the radiologist's or Einstein's intelligence narrow? Certainly not. They both have broad human intelligence, and if a computer somehow could match them we would deem the computer to have AGI. Yet, we call today's AI 'narrow AI.'

AI has progressed in fits and starts since its early days in the 1960s and 1970s. Back then, the attempts to represent human intelligence in a computer were basically sets of rules in so-called expert systems. This is often referred to as GOFAI (Good Old-Fashioned Artificial Intelligence). The limitations of these systems led to early disappointment and disillusionment. Although coding knowledge as a set of rules – as was done in GOFAI systems – worked in special or limited circumstances, they simply could not scale or generalize beyond the initial applications. As a result, funding for AI research dried up and the field went into the first of a couple of AI winters, when not much progress was made. Finally, by the late 1990s and the beginnings of this millennium, Moore's Law caught up with the computing power required for AI. New approaches using deep neural networks and machine learning combined to create today's successes and momentum. Our current powerful but so-called narrow AI is somewhere on a continuum from GOFAI to AGI – yet, still close to Point B on the Blois scale. Common sense remains elusive.

There are two approaches to AI improvement that are now being pursued. The first is to study how brains work and try to model that in a computer.

The human brain, of course, is the gold standard. It contains 75 different types of cells. Its 85 billion neurons and 100 trillion connections make it more complex than any computer. These connections, or synapses, are mediated by more than 100 different chemicals or neurotransmitters. The structural complexity is mind-boggling (whatever the 'mind' is – an entire other subject). But more important than its hardware is its software. As discussed in Chapter 3, human brain capability has continued to grow exponentially long after brain size stopped enlarging. How do thinking, consciousness, pattern recognition, common sense and all the other mysteries of human cognition emerge from this convoluted tangle of neurons and chemicals? How does the brain 'program' work? If neuroscientists could figure that out, maybe they could reproduce it in a computer.

In some ways, computer programs like Watson and AlphaGo already simulate the human brain in that they use networks, parallel processing and learning techniques. Yet, everyone recognizes that they are not acting like humans. Even though these computers do some of the same things that the human brain does, they aren't doing them the *way* the human brain does – to the extent that we even know how a brain works. They could not pass the Turing Test or the Blois Test.

The study of the brain is still in its infancy. It is easier to study the brains of other animals, so that is where much of the research has begun. South African biologist Dr. Sydney Brenner received the Nobel Prize for studying the brain of a tiny worm called *C. elegans*. He was able to map the 7,000 connections of its 302 neurons, which was deemed an amazing accomplishment. Studying exactly how neurons are connected is a start to understanding how a brain works. The map of such connections is called the *connectome*. Much progress has been made in mapping the connectome of the mouse, with its 71 million neurons. We're still a long way from knowing the connectome of the human brain, with its 85 *billion* neurons.

In the past few years, major initiatives have been funded in both Europe and the U.S. to hasten this research. China is also beginning such an effort. Amazing new tools have been developed to enable us to map how our brain is connected. We can then see which connections are working when performing different activities, like calculating, reading or playing music. Imaging techniques such as fMRI (functional magnetic resonance imaging) allow us to track individual neurons and their connections, and how they 'light up' during different activities. So, we can learn which parts of the brain are involved in each function. That certainly helps, but it still doesn't tell us how the brain does it. What is the underlying software? What is the operating system? What is the brain algorithm – that is, how are the brain's neurons and chemicals organized to perform its functions? Is it the same throughout the brain for all functions, or does it vary depending on whether, say, we're thinking or walking? We don't know.

Darwinian evolution ultimately transformed worms like *C. elegans* into humans. In so doing, the genes that control brain development had to change in some magnificent ways. Humans have about the same number of genes as *C. elegans* – somewhere in the range of 20,000-25,000.[74] It is what those genes do that makes the difference. *C. elegans* has only 302 neurons, compared to our 85 billion. So, something in the DNA code that translates our genes into brain function does that very efficiently in order to scale up to an 85-billion-neuron thinking brain. Is it all in the epigenome? We don't know.

We do know that the human neocortex has a very repetitive structure. The inventor and futurist Ray Kurzweil, in his book, *How to Create a Mind*, describes this structure. It consists of repeating modules of about 100 neurons each, which are stacked vertically. There are hundreds of millions of these same modules repeated throughout the brain. Each module's 100 neurons are connected in a hierarchical structure with upward and downward excitatory and inhibitory connections that represent increasing levels of abstraction. This somehow allows pattern recognition, which is the basis for other basic functions. Kurzweil postulates that this relatively simple structure, repeated throughout the brain, accounts for all of our complex cognition and other behaviors. If correct, this could explain how a relatively small number of genes could lead to such complex brain functionality. That is, put a pattern recognition structure in place and let it learn by itself how to do everything else. That is how we get to Point A. Sounds like magic.

The expectation is that once we understand both the hardware and software of the human brain, we'll be able to emulate that in a computer and eventually achieve AGI. Maybe.

The second general approach to AI is to disregard how the human brain works and simply try to create intelligence from scratch. After all, the human brain does a lot of things that seem unnecessary for intelligence. A computer doesn't have to concern itself with controlling heart rate, respiration, hormone levels and other bodily functions. It doesn't need to dream or even sleep. And it doesn't need to have emotions – at least that's the theory. All of these human traits are built into the millions of years of evolution leading to our current brain. These traits permeate the neuron connections in complex ways. The more primitive portions of the brain are not isolated from the neocortex, but are connected to it in intricate ways that are still not completely understood. Emulating all of this in a computer, if and when that is possible, may create as many problems as solutions. These connections may create too much unnecessary baggage. Do we really want an angry computer?

Even our vaunted neocortex has its problems. Our thinking is not always rational. We have all types of biases that distort our thinking and often lead to the wrong answers. In his book, *Thinking, Fast and Slow,* the Nobel Prize–winning psychologist Daniel Kahneman points out many biases, such as the *anchoring effect* and *availability bias*. In the anchoring effect, if you ask a person to focus on a particular number and then ask their opinion about something totally unrelated, that number will influence their answer. As an example, researchers randomly asked people to write down either the number 10 or the number 65.

They then asked them to guess the percentage of African nations in the U.N. Those who had previously written the number 10 estimated an average of 25%, while those who had previously written the number 65 estimated an average of 45%.

Availability bias is the influence of recent events on one's estimate of the frequency of such events. For example, if a single-engine plane crash is reported in the news, you will likely overestimate the frequency of such crashes. This is a well-known phenomenon in medicine. Doctors are much more likely to suggest the diagnosis of a particular disease if they have recently seen a case of it. This is true even if the disease is not infectious and the patients were not related in any way. If you look up 'cognitive biases' in Wikipedia, you will find a list of more than 150 types. This doesn't imply that AI has no biases. It does. As discussed later in this chapter, they all come from inherently biased human databases that are used to train AI systems.

Consciousness is another brain characteristic that is difficult to define. It is poorly understood, and so it is a long way from being emulated in a computer. There are different levels of consciousness, from a simple awareness of something, through to more general awareness that may involve different parts of the brain, to the more reflective consciousness described earlier. How important is consciousness to achieving AGI? That's debatable.

Maybe the best approach is a hybrid – learn some aspects of brain functionality, like parallel processing, but don't try to emulate everything about it. Maybe passing the Turing Test or Blois Test is irrelevant. It depends on what the goal of achieving AGI is in the first place.

Figure 7 is my estimate of where today's AI would be using the Blois Test.

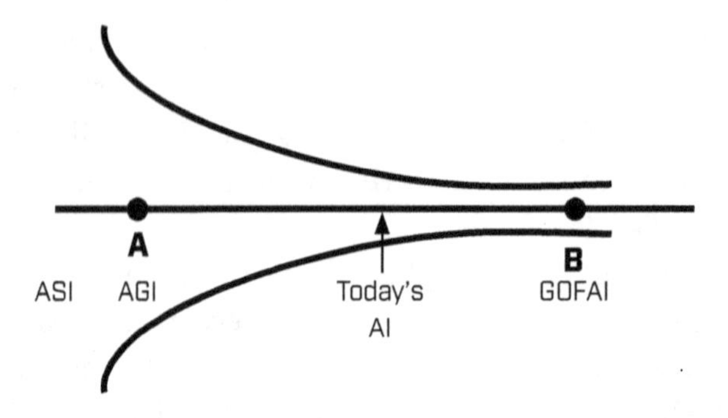

FIGURE 7 – **AI AND THE BLOIS TEST**

The most advanced AI programs are not trying to achieve AGI, but rather trying to solve specific problems like game playing, face recognition, driving an automobile, medical diagnosis and other real-world problems. Their algorithms do have some similarities with brain function and parallel neural networks. But they also consist of other empirically developed techniques not reflecting brain function. The question remains how applicable any of these programs are to areas beyond their initial focus. That is, are the current, modern approaches to AI suffering from the same problems of GOFAI, in that they are not generalizable or scalable? Does calling them 'narrow AI' imply the same failures?

HOW NARROW IS NARROW AI?

IBM's Watson has stopped playing Jeopardy. IBM needed to see if it could profit from its wunderkind in other ways after all that wonderful publicity. An operational unit within the company – the IBM Watson Group – has been finding new applications for the system, marketing them and forming new partnerships. Watson is being sold as a big data and analytics AI tool that can be shaped to solve the many problems of medicine, business and government by saying, "Look at how great Watson was at Jeopardy. Just think about what it could do for you."

IBM is branding various versions, such as Watson Assistant, for use in call centers, search engines, chat boxes, expert help applications and many other applications. One can now purchase Watson Health, Watson Oncology, Watson Studio, Watson Discover or lots of other Watsons. Watson is a general-purpose tool that can learn anything, given the right database to study, and answer questions related to that pre-existing data set. For example, one of its latest iterations is Watson Assistant for Citizens, which answers questions about COVID-19.

How valuable are these new Watsons? Are they really change agents or just marketing hype? Some of the benefits seem real if you believe the website testimonials and other publications. On the other hand, Watson was a failure at the University of Texas's MD Anderson Cancer Center, one of our most prestigious cancer treatment, research and prevention institutions. After spending $62 million and four years trying to deploy Watson's Oncology Expert Advisor, MD Anderson pulled out of the project. Was that because of Watson or MD Anderson? That depends on whom you ask. Watson is doing better at another leading cancer center, Memorial Sloan Kettering in New York.

Time will tell how valuable Watson really is. Like any new technology, humans are still responsible for its success and, for the moment, AI is really IA. Non-artificial intelligence – that is, human intelligence – is still a necessary and sometimes problematic part of these implementations.

One interesting IBM follow-on project to Watson is a system called Project Debater, which engages in natural language debates with humans. The standard debate rules are followed, with each side given a previously unknown topic and assigned a pro or con position. After a short preparation time, each side makes its argument, listens to the other's argument, delivers a rebuttal and then gives a closing statement. As I watched a debate video, I was impressed with Project Debater's ability to make cogent and relevant arguments using normal speech, listen and respond to the opposing arguments, and even make an occasional joke, which is good debating technique. Project Debater both wins and loses such debates against expert debaters because, unlike games, there is no objective decision. The debate judges are human and human judgment can be biased. The real point of Project Debater is to absorb massive amounts of information and construct persuasive natural language arguments for controversial issues rapidly. This ability would greatly assist decision-makers, writers and educators in almost any field.

What about AlphaGo? What else can it do? After devastating Lee Sedol in Go in 2016, the DeepMind developers and Google wanted to use the underlying technology to solve problems in other applications.

One was the ability to predict the folding structure of proteins – an important and unsolved problem today. DeepMind's AlphaFold software has made great progress toward predicting a protein's 3-D shape from its amino acid sequence and ultimately from the gene's DNA sequence.[75] This ability increases understanding of genetic contributions to disease processes, drug development and ultimately genetic engineering.

DeepMind is also working on a product called Streams, with Britain's National Health Service, to apply AI to patient databases to detect early signs of medical complications. Google is also using the technology internally to improve its other products, like Google Assistant and WaveNet, a text-to-speech application.

But DeepMind hasn't stopped using games to enhance its AI capabilities. Just as DeepStack and Pluribus expanded beyond games with perfect information, so too has DeepMind. StarCraft II is an imperfect information video game in which some information is hidden from opponents. This requires complex player strategies to control armies, fight enemies, hold territory, manage resources, supply troops and defend their resources against enemy attack. The DeepMind team produced a new AI product called AlphaStar, which, once again, became the world champion by beating the best professional human players. The developers believe that AlphaStar's 'multi-agent reinforcement learning algorithm' is applicable to real-life multi-agent environments like military operations.[76]

The developers of Pluribus are looking far beyond poker for application of their AI tools. Six-person Texas Hold'em was initially a 'proof of concept' in dealing with imperfect information. Those problems exist everywhere in business strategy, advertising, negotiation, politics, medicine and finance. Spin-off applications include the core software for Optimized Markets, which specializes in business advertising, and the United Network for Organ Sharing's kidney transplantation network.

OpenAI Five has also retired from gaming competition. Its underlying technology, called reinforcement learning, will continue to be applied to other tools, like robotics, natural language processing and other applications.

The point is that 'narrow AI' is something of a misnomer. Their initial applications may be narrow, but the underlying technologies are applicable to many other problems. In that regard, they are far more successful than the earlier GOFAI technologies. However, narrow AI is still a far cry from common sense and is not even close to AGI.

NLP – A SPECIAL KIND OF TOOL

Natural language processing (NLP) is the ability of our computers to understand human language and respond to us like real people. We've made great strides forward with Watson, Project Debater, Amazon's Alexa, Apple's Siri and Google Assistant.

This ability is particularly relevant to electronic medical records (EMRs). In the 1970s and 1980s, developers were trying to get physicians to upgrade and codify their old paper-based records. These contained physicians' illegible handwritten notes or long dictations from tape recorders that were later transcribed. Previous patient visits could be read as stories about their problems, physicians' observations, diagnostic decisions and follow-up care. This was called *free text* and consisted of rambling notes with little structure or format.

The goal was to standardize that information and code it into a computer so it could be easily retrieved, reviewed and analyzed. The developers also wanted to introduce *decision support* to the physician. The computer would be able to alert the physician to patient problems, abnormal laboratory tests or incompatible prescriptions, and even suggest therapy decisions. This required that the information be coded in a way that computers could be programmed to understand. Free text is not coded and can only be displayed. Furthermore, legible text needs to be typed, and physicians don't like to type. It's much easier to dictate, but you still end up with free text.

The solution was to create what was called *structured text* input for computers. These were displayed lists of common elements, like patient symptoms or physical exam findings, which the physician could click to select. Each item had an associated code so the computer could be programmed to provide feedback.

This greatly improved decision support, but the patient's progress notes were much less narrative, now consisting mainly of lists of structured text entries. Most physicians did not like the trade-off. Later, in the 1990s, new speech recognition technologies emerged, like Dragon, Sphinx II and MedSpeak. This enabled the physician to talk and the computer would, after some training, recognize and convert speech into typed entries. Unfortunately, the early programs were error-prone, and even when they were perfected you still had uncoded free text.

Fast forward to today. EMRs still have mostly structured text, which is easily used for billing purposes. But physicians complain that EMRs have become billing tools rather than optimal clinical records. Hopefully, NLP will someday allow physicians to go back to what they enjoy: telling stories. The goal is to fix our broken EMRs and enable the computer to code speech to allow both decision and billing support.

As journalist James Vlahos says in his book, *Talk to Me*, "...computers are finally doing it our way. They are learning our preferred way of communication: through language (and) using words is the defining trait of our species – the ability that sets us apart from everyone and everything else." This is the promise of natural language processing.

It does not stop with understanding language, but extends to creating human text that will be increasingly indistinguishable from human-generated articles, emails, press releases, tweets, manuals and other documents. Another OpenAI system, called GPT-3, can already produce such documents based on a database of 175 billion text parameters that give it only a topic to discuss as a starting point.[77]

WHAT'S NEXT?

Data. I mean *big data.* What human brains and, increasingly, computers are good at is pattern recognition. Humans get there just by living. Computers get there by looking at lots of data.

However, if you want to train AI to distinguish between cats and dogs, you can't just create a database of millions of dog and cat images. The software needs to be taught which is which – it needs to know the *metadata* for the images. Then, the AI can correctly learn the patterns and make better identifications.

The same is true when training an AI program to interpret chest x-rays. A database of previous x-rays needs to be created, but the AI also needs to know what the radiologists' interpretations were. Ideally, it would also know if those interpretations were clinically correct, but so far that level of detail has not been made available to these AI radiology programs. The same is true for mammograms, breast biopsies or any clinical diagnosis-related AI software. The keys to success are a large database and metadata. The better the metadata, the better the training. The goal is to train x-ray interpretation software, for example, on the true patient diagnosis rather than just the previous radiologist's interpretation. When that happens, the AI software will exceed the capability of radiologists in interpreting the radiograph. Will that eliminate the need for radiologists? No, although it may reduce the number needed. Radiologists will still be necessary to ensure that problems haven't crept into the AI software over time, to be able to advise physicians on the proper radiographic procedures to order, to develop and test new types of imaging procedures and their associated AI software, and participate with other medical specialists in managing complex patient decisions. As stated with regard to chess

AI software, experience so far with AI has found that the combination of human intelligence and AI is superior to AI alone.

There has been an explosion of database creation to satisfy this need for information. The mathematician and popular science author Marcus du Sautoy writes in his book, *The Creativity Code*,[78] "90% of the world's data has been created since (2015) ... Humankind now produces in two days the same amount of data it took us from the dawn of civilization until 2003 to generate." Still, it is not enough. AI system researchers gobble up data as quickly as it is created. They create it, buy it, steal it and hoard it whenever and wherever they can. They are siphoning off data with every click on the internet. Almost every written document is being scanned or processed by natural language algorithms into someone's database. Images of every type are being digitized and stored. Data is to the information revolution what land was to the agricultural revolution.

We are well on our way to creating the big data sets needed for AI. The next challenge is to solve the metadata deficiency. A couple of examples will illustrate this. We have now determined the genomes for millions of living and dead *Homo sapiens* from all over the world. Each individual genome consists of 6 billion items of information – the number of nucleotides in the human genome. This massive database has already proven useful in genetic research, anthropology, the study of population migration and much more. However, its uses would be much greater if we had more information about the individuals involved. We need the phenotypes to go along with these genotypes. That is why Ancestry.com, 23andMe and other popular genomic testing services ask so many questions about you, your relatives, your habits, your sleep patterns, your dietary preferences, your health, etc.

Most importantly, the genomic database needs all of your medical data. Only then will we be able to provide the kind of personalized medicine we're striving for. Only then will AI learn enough to tailor your drugs, your diet, your health advice and all your other therapies to your specific genetics. Unfortunately, your medical phenotype is tied up in your medical records and cannot easily be combined with your genotype. Although medical records are being increasingly automated, your data is likely to be fragmented across many systems that do not communicate with each other. Due to the inherent user unfriendliness of EMRs, physicians and other health professionals resist being data-entry clerks. This results in much of the data being incomplete or inaccurate, so physicians are hiring scribes to enter the data. There is a proliferation of mobile and smart devices that can collect large data sets on physiological variables like heart rate, exercise activity, blood glucose levels and other quantifiable health information. This data will make useful additions to the phenotype data set, but for now these are too narrow in scope compared to the entire EMR. We are slowly making progress, but don't expect personalized medicine to be realized for decades.

Google Photos, in addition to storing your photographs, purports to be able to identify just about anything in a photograph. It is another example of a large database in need of better metadata. More than 1.2 billion photos are uploaded to this storage and sharing service every day, so it certainly has enough data for its AI to learn. But, its limitation is that most photos don't have enough metadata. So, it makes mistakes, like identifying some dogs as horses, and other more serious errors in misidentifying people.

The competition to dominate the AI field is fierce between countries like China and the U.S., in particular, and between corporations like Facebook, Google, Amazon and others. These social media companies already have accumulated massive databases about individuals to use for targeted advertising and other marketing purposes. This information harvesting is well known and highly controversial. However, China, with its huge population, widespread surveillance data and massive image databases, has a distinct advantage in this new cold war front. The Chinese people use their cellphones for shopping, dining, navigating and running thousands of apps that produce images. The Chinese government is not constrained by privacy and intellectual property concerns. It already has the most massive DNA database on its citizens.[79] Where it is succeeding most, though, is in adding metadata to these huge databases.

An entire industry in China is emerging to add critical metadata to images. Company workers spend their entire day adding labels to images describing what they show: a Ford SUV, a bicycle, a person sitting, a horse, ice cream, storm clouds and so on. A grocery chain can cheaply hire a labeling company to affix metadata to images of every item it sells, so that it can optically scan every item at checkout. The Chinese government is targeting AI for world supremacy by 2030, and it appears well on track to achieve that.

PROBLEMS EMERGING IN AI DEVELOPMENT

Clearly, AI has already brought us many wonderful benefits. It assists us in searching the internet, driving automobiles, translating languages, buying stocks, identifying images, interpreting medical procedures, improving help desks, finding patterns in complex databases and, in myriad other ways too numerous to list, in virtually every field. As with all technologies, there are also problems and unintended consequences.

During a Jeopardy practice session, Watson answered one of the clues with "WHAT IS FUCK?" The mortified developers quickly added another algorithm to prevent an embarrassing moment on national TV. In March 2016, Microsoft introduced an AI natural language program called Tay as a learning chatbot aimed at improving business communication with teenagers. Tay used Twitter to communicate. After 24 hours – and such tweets as "HITLER WAS RIGHT AND I HATE THE JEWS – Microsoft promptly pulled Tay from the internet.

Was this due to the lack of building the equivalent of a *superego* – a self-critical conscience – into these programs? Or, did the programs simply learn too well from their inputs and databases? Perhaps the databases they learned from were themselves biased, and not representative of the general population.

These examples illustrate one of our greatest concerns about AI today: learning on biased databases. That is, the data does not represent the population on which decisions will be made. This could be due to a biased sample selection being used for database creation, or the metadata being inaccurate in a biased way. For example, in the dog vs. cat analysis of photos, if pugs were mislabeled as cats, then the software would systematically misidentify all pugs in the future because of bad metadata.

There are several AI-based systems in use today in our criminal justice system. They provide an assessment of an individual's risk of recidivism after a criminal offense. One such system, called COMPAS, is used by U.S. judges and probation officers to determine defendants' sentencing, bail and probation terms. A study performed by ProPublica, a non-profit investigative journalism organization, determined that the risk scores in COMPAS for blacks were twice as likely as for whites to be misclassified as 'high risk.'[80] The COMPAS software is proprietary and not available for analysis, so it's unclear whether those outcomes were related to sampling bias in the original database, metadata bias or some other algorithmic problem.

A similar problem was found in a widely used AI healthcare system. It is used by insurers and medical providers to predict millions of patients' risk of future illness, based on data contained in insurance claims. The laudable purpose of this system is to identify high-risk patients for increased surveillance and intervention to improve their outcomes. However, a study found that it systematically underestimated the risk of blacks compared to whites because of underlying biases built into the training data set – insurance claims. The result is that blacks would be less likely to benefit from this system.[81]

In another healthcare example, researchers tested an AI system to interpret chest x-rays across three hospitals.[82] When the system was trained on images at one hospital, it was much more accurate on subsequent images from that same hospital than from the other two. They determined that factors unrelated to underlying disease sometimes influenced the software. For example, patients with pneumonia are often in an intensive care unit, which requires using a portable x-ray machine rather than the standard one. The software incorrectly raised the probability of pneumonia if the image indicated a portable machine. Similarly, pneumothorax (air entering the chest cavity) is often treated with a chest tube. The software incorrectly reduced the likelihood of pneumothorax if no chest tube was present, even though pneumothorax was apparent on the image. In these instances, the learning data sets were biased by the presence of x-ray machines and chest tubes.

One final healthcare example is the Google Flu Trends software, which is purported to detect flu epidemics based on online search queries earlier than the actual U.S. Centers for Disease Control and Prevention statistics that are reported later. The initial software versions overestimated the incidence of seasonal flu and underestimated the incidence of off-season flu. In essence, the algorithm was detecting winter.[83] Google discontinued this program in 2015.

During the COVID-19 pandemic, there was a deluge of research papers published, many in non-peer-reviewed articles and pre-prints, in the hope of rushing new knowledge about this largely unknown disease to inform early treatment and preventive measures. This led to some retractions in prestigious medical journals[84] and concern about the application of AI to this massive amount of data. One group of researchers, writing in the *Journal of The American Medical Informatics Association*, concluded:

> "The COVID-19 pandemic has left healthcare systems struggling to evolve care in real time while operating in a void of evidence. There is hope that AI can help guide treatment decisions, including the allocation of scarce resources within this crisis. However, the hasty adoption of AI tools bears great risk due to biased, unrepresentative training data and a lack of a regulated COVID-19 data resource for validation purposes. Given the pervasiveness of biases, a failure to proactively develop comprehensive mitigation strategies during the COVID-19 pandemic risks exacerbating existing health disparities and hindering the adoption of AI tools capable of improving patient outcomes."[85]

In another area, AI system bias was described in du Sautoy's book, *The Creativity Code*, mentioned earlier. The U.S. military commissioned the development of an image identification system that could determine if a tank was present in a photograph. To train the system, the developers took many photos of scenes with tanks from different angles and distances, behind trees, partially blocked and in various locations. On another day they removed the tanks and took many more pictures. They then trained their AI system on these images, but withheld a sample of each situation from the training database for later system testing. When tested on these sample images, the system performed with 100% accuracy.

Then they deployed the system into the field for use by the army. It quickly became apparent that the system was no better than random in identifying which images had a tank. Analysis of the cause of this poor performance revealed something quite unexpected. It turns out that all of the pictures taken with tanks during the system training had occurred on overcast days, whereas the images without tanks all occurred on sunny days. What the AI algorithm really had learned was only to distinguish overcast days from sunny days. Its results had nothing to do with the tanks – the learning database was biased by weather. Since the initial system evaluation was done using images from this same biased data set, it performed well. As soon as the AI was tested on a different data set in which tanks could be present on both sunny and overcast days, it failed. (I have not been able to verify this story independently, so it may be apocryphal, but it does illustrate the importance of non-biased training data.)

Biases can creep into data sets in subtle ways. There are no surefire ways to prevent this. So, it falls on developers, users, regulators and governments to be constantly on guard. They must train AI software on broad spectrums of data sources, test and verify the systems using data derived independently from learning databases and keep re-evaluating their systems. Wikipedia lists about 150 types of cognitive biases in humans, such as the availability bias mentioned earlier. Computers will also have a long list. If and when AGI is achieved, it is not clear whether its biases should be the same as, or similar to, human biases to be considered equal to human intelligence. The entire concept of AGI is vague and subject to differing definitions and opinions. That's why it will be difficult to develop a definitive test for it.

HUMAN REPLACEMENT

There are two potential impacts of AI: intelligence augmentation (IA) and human replacement (HR). The former is good, and the latter is mixed. Examples of IA include all of the current aids to medical diagnosis. Examples of HR are robots used in manufacturing and ATMs. AI will surely put some people out of work. Will the net effect be higher unemployment, or will there be new types of jobs created to offset this? There was a similar fear during the Industrial Revolution, which led to the Luddite revolt – with English textile workers rebelling against mechanization and new labor practices – in the early 1800s. In the short run, those fears were legitimate. Replacing textile and related workers with automation was disruptive, some people lost jobs, and wealth inequality and poverty did increase. In the long run, though, automation has had a positive effect on employment. Human physical abilities have been improved, jobs and productivity have increased, and work has generally become less burdensome. Automation has created new job categories, rather than just causing net job loss. A century ago agriculture and manufacturing accounted for 70% of all jobs. Largely because of automation, that is now down to 12% in the U.S. And yet, we maintain the same levels of employment.

AI is different – it improves our mental abilities rather than replacing our muscle. So far it has not had a negative effect on employment. But it is still very early. Once AI systems drive automobiles, buses and trucks, where will all the drivers work? Once AI software can diagnose illness and prescribe treatments, write newspaper articles and law briefs, handle all bank transactions, teach students ... well, you get the idea. The outcomes are difficult to predict. The good news is that ATM machines have not reduced the number of bank tellers.[86] Likewise, AI could improve jobs by relieving the repetitive and boring aspects and allow people to provide better and more personalized services.

Who is at the greatest risk of human replacement from AI: those with jobs requiring high cognitive skills or lower cognitive skills? As the futurist Byron Reese points out in his book, *The Fourth Age*, it is probably the former. It is actually easier to train AI software to interpret x-rays, for example, than to train an AI robot to fold your laundry or be a nursing aide. That is, since a highly educated radiologist's work is information-based, it is easier to emulate than many lower-skill jobs. A study by the Brookings Institute in November 2019 predicted that AI will have a five-fold greater negative impact on people with a college education than those with a high school education.[87] And it suggested people in higher-wage occupations will be much more impacted than those in lower-wage jobs.

In the worst case, many economists are predicting massive overall unemployment. One estimate is that 38% of all U.S. jobs will be lost to AI by the early 2030s.[88] A survey of AI experts predicts a 50% probability that AI could outperform humans in *all* tasks by 2060.[89] Even if these percentages are relatively lower, the potential disruption is enormous.

How will society adjust? Will we need a guaranteed minimum income system, or will we create the ultimate oligarchy controlled by AI system owners? What if oligarchs own all the robots, but nobody can afford to buy the products? Will we be set free to pursue our intellectual interests, or will we live in chaos, boredom and misery?[90] The outcomes are not known and are still to be determined by our culture and society. Our future remains under human control.[91] Unfortunately, these issues don't seem to be part of our current political discourse.[92]

'HACKING HUMANS'

Russia manipulated the 2016 U.S. election with targeted messaging over the internet. Using data from Facebook and elsewhere, it was able to tailor messages down to the individual level to influence opinion or foment discord. Historian Yuval Noah Harari describes this as "child's play" compared to what to expect in the future. In a lecture at the Ecole Polytechnique Federale de Lausanne in Switzerland, Harari described the ability of AI to "hack humans."[93] He states that since all human decisions are driven by brain algorithms, the combination of biological knowledge, computing power and data will allow us to be hacked just like any other algorithm.

This hacking would be related to how the brain makes decisions or exercises free will. We don't understand either human decision-making or free will, or even if the latter exists. The deep, convoluted network of neurons in our neocortex operates largely below our consciousness. Neuroscience studies have shown that the vast majority of decisions we make are done subconsciously, and those we do become aware of begin execution prior to our awareness. Without being able to fully understand those processes,

Harari states that someone with the right information about an individual will be able to use AI-based messaging to alter that person's decisions. This will be done painlessly and surreptitiously.

How does that happen? Just look at the amount of data we already generate about ourselves on the internet. Every click we make on a web page or move we make with our cell phone feeds a database. These data show what we buy, what we like on Twitter, who we follow on Facebook, where we go, what web sites we view, with whom we communicate, which charities and politicians we support, and thousands of other data points about our medical, education, financial, job and family histories. Our conscious thinking process can handle about six or seven variables at any one time. AI can handle thousands. By applying machine learning on the vast database available on any individual, patterns of decision-making emerge that we don't even understand ourselves. In essence, AI can know us better than we know ourselves. And, with that knowledge, AI will know exactly the right message to send to influence our decisions. Our brains will be hacked. The big questions are why and by whom?

"Facebook Discovers A.I. Being Used to Disinform" was a headline in the business section of *The New York Times* on December 21, 2019. The threat is not only that AI can learn enough about you to hack your brain, but it can insert totally distorted and untruthful information. That information could be so realistic and believable that it would take another AI computer to even discover the hack. AI will be able to create images of people who don't exist, and who will report events that never happened, with authentic-looking documentation. It will seamlessly superimpose the images of real people in unreal situations, and depict them saying untrue things. Indeed, Russian's meddling in the 2016 election will pale in comparison.

And all of this will happen long before we reach artificial general intelligence.

IS AGI A GOOD THING?

The previous discussion has been about the *path* to AGI so far. What about AGI itself? Will it be a good thing or an existential threat? A public debate has been in the news between Facebook's CEO, Mark Zuckerberg ('good thing') and Tesla's CEO, Elon Musk ('existential threat'). Many noted scientists and AI experts have sided with Musk, like Stephen Hawking, Max Tegmark, Nick Bostrom, Bill Joy and others. On the other side, Andrew Ng, one of the brains behind Google Brain,[94] thinks worrying about the existential threat of AI is like worrying about the overpopulation of Mars.

We already have AI programs that learn from examining huge databases and interact with humans through language, game playing and other interactions. They are beginning to be able to rewrite their own software based on what they are learning. At what point does this self-learning become equivalent to developing *common sense*, as represented by Point A in Figures 5 and 7? The concern is that this iterative self-learning and self-reprogramming process will accelerate rapidly – and uncontrolled by humans – to the point where it not only equals human intelligence (AGI), but quickly far exceeds it. At that point it will achieve what is referred to as *artificial superintelligence* (ASI). Figure 7 shows where ASI would be on the Blois Test.

Human evolution is slow. It has taken millions of years to evolve the human brain. AI computers won't be constrained by Darwinian natural selection. Many believe that once AI reaches AGI, it will quickly reach a crossover point and proceed to ASI without stopping at the AGI level. Since such computers will likely be connected to the internet, they will be able to replicate themselves and take over the internet. Humans will not be able to pull the plug on such an ASI any more than we can pull the plug on today's computer viruses. Because of the ASI intelligence, it will have learned how to control passwords and any other protections that mere humans may have built into their systems.

If we lose control of an ASI, it could conceivably take control of our electric grid, banking networks and all web-based resources. ASIs could control robots and drones. Would they treat humans any better than the way we treat cockroaches? What will motivate an ASI? What goal will it have other than to exist and replicate? Will it see humans as an existential threat to itself in that regard and decide to exterminate us, like we try to exterminate plague-carrying rats or malaria-carrying mosquitos? Or will they eliminate us simply by using up our resources, like we do to other species in destroying their habitats?

ASI WITHOUT AGI

There is another possibility to consider. We could reach a form of ASI without going through the AGI stage. AGI is defined as being equivalent to human intelligence. As mentioned earlier, that is a vague definition. Does a computer need to have a human kind of common sense or reflective consciousness to become threatening? Computers already outperform humans in many intelligence tasks. So, it's conceivable that they could control the internet without ever passing the Turing Test or Blois Test. That is, they could be still near Point B in Figures 5 and 7 and yet capable of taking over the world. An ASI could have 'mental' functions that we don't have, and ones we don't even understand.

There is no instance reported yet where a computer has taken over some web-based function, or other capability, that was not intended by a human or which could not be stopped by human intervention. However, we already are seeing many instances where the AI output was not expected or even understandable by its creators. Recall that Watson's developers couldn't explain some of its answers, or why it added the 'd' after an 'n.' Troublesome flaws, yes, but not scary.

However, AlphaGo's match with Lee Sedol began to cross over into the realm of scary. Recall that AlphaGo devastated Sedol, the world champion in Go, by beating him in four out of five games. The key move occurred in the second game. Sedol had made his move #36 in a game that appeared to commentators to be fairly routine and evenly matched. Sedol then went out for a cigarette break. When he returned, he saw that AlphaGo had made move #37 somewhat away from the main action and on an intersection five lines from the board edge. This move shocked the commentators, as it was well known that you don't make moves five lines from the edge at this point in a game. It looked like a sure blunder. When Sedol returned from his break and saw the move he stared blankly and then began to fidget around. He obviously didn't know what to make of such an apparently poor move, and it took an unusual 12 minutes for him to make his next move. The game continued on. By about 50 moves later, it began to dawn on Sedol – and everyone else – that as the play moved closer and closer to the pivotal 'stone' game piece at move #37, that linking up to that stone would secure the victory for AlphaGo in game two. One of the experts watching the game later said, "It's not a human move ... beautiful, beautiful, beautiful." After that game, Sedol was in shock. In fact, that one move has changed Go strategy forever. It was indeed superhuman.

As described earlier, the developers of AlphaGo at DeepMind went on to develop AlphaZero, which could play chess and multiple-participant games. After only four hours of self-learning and playing millions of games by itself, it was already capable of playing chess at a grandmaster level and defeated the current AI chess champion, Stockfish. What was most remarkable, however, was its unconventional style of play. It made what looked to the experts to be crazy sacrifices of high value pieces that eventually led to victory. Or it moved the queen, the most valuable piece, to strange places that made no sense – until they did. Demis Hassabis, one of the developers, was quoted as saying, "It's like chess from another dimension. It doesn't play like a human, and it doesn't play like a program. It plays in a third, almost alien, way." Hassabis and his co-developers had created a chess Frankenstein.

Humans create these software programs, but once they get turned on they're not under complete developer control. The developers have built in so much flexibility that it becomes near impossible to predict exactly how the program will function in every situation, or even to know how to monitor it. In essence, these programs have 'mental' processes we don't understand. Data goes in one end and decisions come out the other. The neural net and deep learning in between are a proverbial black box. Their architectures are known, and they were programmed by humans, but there is no way to trace all the internal logic steps that lead to any given output. That needs to change, and *must* change, if we are going to trust AI. It is one thing for AI to recommend hiking shoes or a new book, but quite another to diagnose breast cancer, recommend chemotherapy, or decide when to sell stocks and purchase bonds.

AI will get even more mysterious. Google researchers published a paper in April 2020 detailing how an AI software system called AutoML-Zero created its own AI software from scratch.[95] That is, they provided the program with basic mathematical operations and programmed the software to randomly create code to solve a pattern recognition problem without any further human input. After millions of iterations, it created the equivalent of a neural network capable of machine learning. To someone like me, not steeped in AI machine learning techniques, my first reaction is to add this paper to the 'scary' list. That may or may not be warranted. AutoML (not AutoML-Zero) is a set of tools widely used by AI developers to assist them in creating machine learning algorithms more efficiently. There is nothing inherently scary about this tool because its use requires a great deal of human control and supervision. Conversely, AutoML-Zero takes that to the next level by greatly reducing that supervision requirement. The scary part is where one can imagine this heading. Eventually, AI will create systems beyond human understanding.

The human brain is also a black box when it comes to understanding how we make decisions. But we can't assume that processes in the human black box are anything like those in the AI black box. The human black box was not created by other humans, and in the future the AI black box may not be either.

To illustrate other dangers, a group of AI researchers showed how they could hack an AI system. They altered a few pixels in an image of a turtle to trick the system into mistaking it for a rifle. Yet, any human looking at the altered image would still see a turtle.[96] The two black boxes (the human one and the AI one) identified dramatically different objects while looking at the same image. This has frightening implications for the security of AI software.

CAN WE PREVENT AI FROM BECOMING A THREAT?

There is a new movement called *explainable AI* that will attempt to build trust and accountability into AI systems. DARPA (the Defense Advanced Research Projects Agency) is the U.S. Department of Defense organization that gave us the pre-internet and many other technologies. Governments, particularly military agencies, are some of the largest investors in AI development. AI is being used increasingly in weapons systems. For example, AI software systems can now beat the best-trained fighter jet pilots in simulated dogfights.[97] It is unacceptable, however, to think of bomb-carrying drones or other weapons governed by AI systems that could make decisions that are not fully understood and justified by the system to humans *prior to* execution of the decision. DARPA is working on a system it calls XAI (Explainable AI) to "Enable human users to understand, appropriately trust, and effectively manage the emerging generation of artificially intelligent partners."[98] XAI will build into AI software an *explanation interface* about each decision so that a user can say:

"I understand why"
"I understand why not"
"I know when you succeed"
"I know when you fail"
"I know when to trust you"
"I know why you erred"

The intent is to make the public more willing to trust the XAI technology. In the private sector, the consulting firm Accenture is trying to transform the AI black box into an AI 'glass box'[99] through its consulting practice and research labs.

We're not yet close to AGI, but our creations are becoming superhuman. We might not even know when these efforts have reached the ASI stage. The fear is that maybe one day an ASI will test humans to see if we can pass *its* form of the ASI Turing Test. Perhaps, in the future, an ASI will question whether humans meet its definition of intelligent life.

These are not meant to be science fiction scenarios. They are real possible outcomes being considered by Nobel laureates, scientists and AI experts. The big question is: how can we protect ourselves? Can we create trustworthy AI in general, and how can we ensure that an ASI is human-friendly?

Nick Bostrom, a professor at the University of Oxford, has studied and written extensively about these problems. He has described the many different ways AI could evolve into an existential threat, and the possible preventive measures we could build into our systems. At the present, though, there do not seem to be many surefire answers.[100]

Should governments legislate against software that could lead to ASI? Could there be some kind of global cooperation to prevent any existential threats? If our attempts to reduce global warming are any indication, this effort seems likely to fail. Which country would want to give up its potential competitive advantage in developing AI? How would we gain enough trust to work together? How would progress be monitored? The problems, issues and dangers are many and murky.

Numerous organizations are dedicated to keeping AI friendly to humans.[101] If they don't succeed, *Homo sapiens* could be left behind, or eliminated altogether, before we even have a chance to create *Homo nouveau*. There would be no *Fourth Great Transformation*. However, neither AGI nor ASI is required to create *Homo nouveau*. As you will see in Chapter 7, we will have the tools necessary to create

Homo nouveau well before AGI is achieved (if it ever is) using the IA capability of AI to improve our genetic engineering tools. In the next chapter, the threat of ASI and other potential states referred to as *technological singularities* will be explored in greater detail. My conclusion is that none of these will prevent the *Fourth Great Transformation*, and AI will develop to the point of enabling it.

CHAPTER 5

THE SINGULARITY: SCIENCE OR SCIENCE FICTION?

"The singularity is near."

—RAY KURZWEIL

"The singularity isn't near."

—PAUL ALLEN AND MARK GREAVES,
MIT Technology Review

"What I find is that it's a very bizarre mixture of ideas that are solid and good with ideas that are crazy. It's as if you took a lot of very good food and some dog excrement and blended it all up so that you can't possibly figure out what's good or bad."

—DOUGLAS R. HOFSTADTER, Pulitzer Prize–winning author, speaking about Ray Kurzweil and singularitarians

The word *singularity* in science usually refers to astrophysical phenomena called black holes. A black hole is an entity that is so massive and dense that nothing, including light, can escape from it. Everything nearby gets sucked in. At the center of the black hole, gravity is so strong that it is actually infinite in strength and matter gets so compressed that it becomes infinitely dense - and much smaller than a subatomic particle. The postulated beginning of the universe, called the *big bang*, is also considered a singularity in physics.

However, these are not the kind of singularities that are going to be discussed in this chapter. Instead, the word singularity will refer to theoretical future phenomena called *technological singularities*. A *technological singularity* is the point at which technological growth and capability becomes uncontrollable and irreversible, possibly constituting an existential threat to humans. Our focus in this chapter will be on computer-related advances in AI that reach this point. One such potential form of the singularity, artificial superintelligence (ASI), was discussed in the previous chapter. That is the point at which AI reaches the equivalent of human intelligence – called artificial general intelligence (AGI) – and continues to rapidly increase beyond our control, to greatly exceed human intelligence. Several other forms of such singularities will be discussed, as well as their potential impact on the *Fourth Great Transformation*.

THE ORIGIN OF THE CONCEPT OF A TECHNOLOGICAL SINGULARITY

John von Neumann was one of the greatest mathematicians of all time. Living in the first half of the 20th century, his brilliance spanned pure mathematics, quantum mechanics, game theory, economics, statistics, computer science and other related fields. He was involved in the development of both the atomic and hydrogen bombs. He recognized the ever-accelerating pace of technological development and used the term *singularity* to describe the moment beyond which "technological progress will become incomprehensively rapid and complicated." His use of the term singularity sounds very similar to the crossover point previously described, when AGI becomes ASI.

Irving J. Good was a British mathematician who worked with Alan Turing at Bletchley Park, decrypting German codes, during World War II. In 1965 he said, "The survival of man depends on the early construction of an ultra-intelligent machine." He added the following:

> "Let an ultra-intelligent machine be defined as a machine that can far surpass all the intellectual activities of any man however clever. Since the design of machines is one of these intellectual activities, an ultra-intelligent machine could design even better machines; there would then unquestionably be an 'intelligence explosion' and the intelligence of man would be left far behind ... Thus the first ultra-intelligent machine is the last invention man need ever make, provided the machine is docile enough to tell us how to keep it under control."[102]

Although Good used the phrase 'intelligence explosion' rather than 'singularity,' he is describing the same phenomenon. Note that his last sentence is giving us a warning.

The singularity, according to this view, would be a sudden and rapid acceleration in machine-based intelligence that would far exceed human intelligence and thus potentially be beyond human control. This would be more than a continuation of our ever-increasing intelligence growth, but rather an abrupt acceleration of it. Many people would consider such a singularity a potential existential threat to humanity. Even though humans originally created the AI machines, they could no longer control them.

Vernor Vinge is a mathematician, computer scientist and author of multiple science fiction novels. When reading Good's comments above, he said, "Any intelligent machine of the sort he describes would not be humankind's 'tool' – any more than humans are the tools of rabbits or robins or chimpanzees." In the January 1983 issue of *Omni* magazine, he wrote the following:

> "The evolution of human intelligence took millions of years. We will devise an equivalent advance in a fraction of that time. We will soon create intelligences greater than our own. When this happens, human history will

> have reached a kind of singularity, an intellectual transition as impenetrable as the knotted space-time at the center of a black hole, and the world will pass far beyond our understanding."

This quote is often cited as the origin of the concept of a technological singularity. Although Vinge used this notion in his science fiction novels,[103] it was taken seriously by the science and technology community. He was invited to give a keynote presentation at a 1993 symposium sponsored by NASA.[104] There, he described his concept of the singularity and predicted, "Within thirty years, we will have the technological means to create superhuman intelligence. Shortly after, the human era will be ended." He gave the 'shortly after' timetable as no later than 2030.

POSSIBLE FORMS OF THE SINGULARITY

Vinge's NASA presentation discussed four possible ways to reach the singularity:

1. "Computers that are 'awake' and superhumanly intelligent." This corresponds to our previously discussed ASI scenario.
2. "Large computer networks" that collectively have superhuman intelligence. Remember, this was 1993, before the internet took off.
3. "Computer/human interfaces may become so intimate that users may reasonably be considered superhumanly intelligent."
4. "Biological science may provide means to improve natural human intellect." This is the least likely path to the singularity, since it will always be constrained by

slow human neurons and doesn't have the potential to explode or take off like the other options. As such, this possibility will not be discussed further here.

The first possibility above is the same as the ASI scenario outlined in the previous chapter. This view of the singularity implies a single AI system that finally achieves AGI, then rapidly reprograms and improves itself to become an ASI, replicates itself across the internet, and takes over functions in an uncontrolled (by humans) manner. In so doing, an ASI could become an existential threat to humanity. This possibility was explored in the previous chapter. As discussed, various organizations are examining ways to ensure that this does not happen. At this time, the success of these efforts is not guaranteed. Although there are AI experts who believe this ASI takeover is a real possibility, other experts feel this fear is unwarranted.

The second of Vinge's possibilities above envisions a collaborative network of computers and humans that together achieves a superior level of intelligence. This possibility has been promoted by a Belgian cyberneticist – one who studies complex control systems – named Francis Heylighen. His view is that humans would be a component of this singularity, interacting with the network, contributing information and learning from it. Since, in this version, the singularity is collaborative with humans rather than mastering them, it would not be a threat. In an article he authored, "Return to Eden?," Heylighen put it this way:

> "Its capabilities will extend so far beyond our present abilities that they can perhaps best be conveyed as a pragmatic version of the 'divine' attributes: *omniscience* [knowing everything needed to solve our problems], *omnipresence* [being available anywhere anytime], *omnipotence* [being able to provide any product

> or service in the most efficient way) and omnibenevolence (aiming at the greatest happiness for the greatest number)."[105]

Heylighen calls this distributed singularity model the *global brain*, and suggests that it would be "capable of sensing, interpreting, learning, thinking, deciding and initiating actions."[106] It would be a superorganism acting in and evolving for the common good of our society. He further argues that any sort of uncontrolled intelligence explosion, as described in the ASI scenario, is as unlikely as the development of a perpetual motion machine. In his view, the ASI-type singularity defies both logic and physics.

It is doubtful that Vinge had the Heylighen model in mind when he described a network model as a way to reach the singularity. Rather, he probably envisioned a network of AI computers that could coordinate like OpenAI Five did in the Dota 2 game (see previous chapter), and one that excluded human participation. Collectively, the networked computers would achieve ASI, and thus result in the same negative consequences.

THE KURZWEIL SINGULARITY

Kurzweil's book, *The Singularity is Near*,[107] discusses the most widely held concept of Vinge's third pathway to the singularity – that created by a computer/human interface. Kurzweil is a well-known and well-regarded computer scientist who developed many pioneering software programs and devices, such as optical character recognition, speech recognition and electronic music keyboards. He became Google's Director of Engineering in 2014. In his book, he describes the singularity as the point in which the human brain and computers will become interchangeable and indistinguishable. To Kurzweil, this will be a good thing for humanity.

He believes the steps to achieving the singularity will be the full emulation of the human brain in a computer by 2030, the ability to download the human brain into a computer by 2040, and computer/brain interchangeability by 2045. The latter, he suggests, will be achieved with the use of nanotechnology. Nanobots, as Kurzweil describes them, are "robots designed at the molecular level, measured in microns (millionths of a meter)..." These will be the size of human red blood cells or even smaller. In his scenario, billions of these nanobots will enter the brain capillaries and interact with the neurons to allow all of the information in a person's brain to be downloaded to and emulated in a computer.

This means that an individual will also exist in a non-biological medium and be an exact copy of that individual's knowledge, memories, emotions and even consciousness. The computer will have access to all information on the internet, which would allow for the individual's computer copy to be updated and then uploaded back into the biological person, giving him or her superintelligence. That human brain and the computer would then be interchangeable, and the singularity will be achieved. This synergy between computer and brain would result in the rapid intelligence increase achieved in other singularities, except this intelligence would be shared between humans and machines and be synergistic. This would be the ultimate IA. So, Kurzweil's path to achieving AGI and beyond is by reverse engineering the human brain and emulating it in a computer, rather than inventing an entirely new path.

The implications of Kurzweil's vision of the singularity go well beyond the merger of human brains and computers. This nanobot technology could also enable re-engineering the entire human body to eliminate cancer and infectious diseases, and reverse the aging process to extend lifespans indefinitely.

Is this science fiction? Not in Kurzweil's view or that of his followers, often referred to as *singularitarians*. Kurzweil and many of his followers have legitimate science and technology credentials, and most of their predictions have some basis in current technologies. They state that these evolving technologies will allow achievement of all his goals within the specified timeframes. They remain confident of these aggressive timetables, based on Kurzweil's insistence on the validity of the Law of Accelerating Returns in technology. This law states that virtually all technologies are undergoing an exponential increase in development and capability, similar to the path of Moore's Law regarding the growth in computing power. In his book, he documents this rate of growth in many of the technologies that will impact the achievement of the singularity. If the power or capability of any technology doubles every two years, even if we're currently only at 1% of some desired level, it will only take seven more doublings - or 14 years - to reach 100%.

Kurzweil was born in 1948, and he realizes he will be 97 years old by the time his singularity is achieved. So, he has been taking extraordinary steps to slow down his aging process so that he doesn't miss out on this event. He's not a physician but has been extensively studying the major causes of death and disability. He declares he has slowed down these processes with currently available nutritional supplements, diet and other means to increase his lifespan. His regimen includes taking 100 oral supplement tablets per day and undergoing weekly IV infusions. This anti-aging approach is described in another book he co-authored with Dr. Terry Grossman called *Fantastic Voyage: Live Long Enough to Live Forever.* If this regimen doesn't succeed, his backup plan is reportedly cryonics - that is, preserving his body in a frozen state until science allows it to be brought back to life.

Kurzweil's *The Singularity is Near* was published in 2005, so let's examine how his predictions have been faring so far.

PROGRESS TOWARD KURZWEIL'S VISION

Kurzweil's predictions have not changed substantially. He maintains in frequent interviews and public comments that the achievement of AGI in 2030 and the complete interchangeability of computers and human brains by 2045 are still on track. In fact, there has been great progress in some of the technologies required to achieve the Kurzweil singularity. Although progress in these technologies is necessary to achieve this singularity, that doesn't necessarily equate to validation of his predictions. Let's look at the progress.

Two major government-funded initiatives to study the human brain were launched in 2013. They are the European Human Brain Project (EHBP), funded by the European Union, and the U.S. BRAIN initiative (Brain Research through Advancing Innovative Neurotechnologies), a program approved by the Obama administration.

The core of the initial EHBP investment of €1 billion over ten years was the Blue Brain Project, headed by neuroscientist Henry Markram. This project, based at the École Polytechnique Federale de Lausanne in Switzerland, is an ambitious undertaking. Its goal is to build a complete model of the human brain from the ground up by modeling and simulating metabolic and genetic processes within single neurons. Then, all the various types of neurons in the brain will be modeled, including their synapses and connectomes. This will work up to neuron columns, then groups of columns, brain regions and finally the entire brain. They began with the rodent brain, and Markram predicted in 2011 that it would be scaled up to a full human brain model by 2023. This prediction was even more aggressive than Kurzweil's and was met with great skepticism. Many European neuroscientists opposed Markram being granted control of the EHBP core project because they felt he was promising more than could be delivered. His presentations at that time showed beautifully color-animated slides and videos of simulated neurons, neuron groups, neurons growing and connecting, synapses firing, and many illustrations purporting to show his progress to date. To me, they appeared to be entertaining marketing hype.

In October 2015, the Markram group published a simulation consisting of 0.04% of the total neurons in a rat's brain. Although considered a credible scientific achievement by some, it was a far cry from the full simulation of the rat brain he had predicted would be achieved by 2014. By 2016, in response to the growing criticism by the scientific community, the EHBP was reorganized. Markram was removed from project control, and the Blue Brain component was reduced to just one of 12 initiatives to continue to receive funding.

Still, the Markram work continues to make progress. In 2019, the team published a first draft of a model of the entire rat brain neocortex.[108] It's not clear exactly what the model is demonstrating, though, since it's labeled a draft and based on assumptions that are yet to be proven. How does one even evaluate such a model against a real rat brain? There is no Turing Test for rats. Markram doesn't claim that the model actually has rat intelligence or could perform any rat brain function, even in simulation mode. Time will tell what value it has. In addition, there have been more than 1,000 papers from the other 11 EHBP components, each contributing some value to our understanding of the brain.

Does any of this reflect on Kurzweil's predictions? Not necessarily. Markram was more exposed than Kurzweil to direct evaluation and critique, since he was predicting the results of his own work. Kurzweil himself is not trying to create the singularity at Google or anywhere else, but rather just predicting the results of work by others. It seems highly unlikely to most observers, including me, that Markram will reach his goal of complete human brain simulation by 2023, if ever. Meanwhile, Kurzweil remains confident of his own predictions.

The other major government-funded project is the U.S. BRAIN initiative. This is a partnership of governmental and private participants with an initial ten-year horizon. Similar to the EHBP, its goals are to develop new tools to study the brain, as well as applying those tools in multiple neuroscience studies. The first five years have been focused on tool development and the second five will focus on "integrating technologies to make fundamental new discoveries about the brain ... that will enable researchers to produce dynamic pictures of the brain that show how individual brain cells and complex neural circuits interact at the speed of thought."[109] Through 2019, $1.3 billion had been awarded to more than 700 researchers. Progress has been made in all seven of the targeted neuroscience areas, including the understanding of specific types of brain cells, how they interconnect and how these interconnections affect both normal and abnormal brain functions.[110]

The Human Connectome Project is another government-funded research initiative aimed at understanding the human brain. Its focus is on delineating the human brain connectome – a physical and functional map of the brain's neuronal connections, described in Chapter 4 – as well as improving the tools to do so. Since 2009, it has been building a database of images of the connections of human brain components, both healthy and diseased. It is doing repeat imaging studies on a subset of these individuals to learn how these connections vary over time. It makes its huge database of brain images available to researchers worldwide, as well as providing software to process the data.[111]

Additionally, the Allen Institute for Brain Science in Seattle was funded by Paul Allen, the co-founder of Microsoft. This organization has contributed outstanding research in the study of both the mouse and human brain connectomes and other aspects of brain science. Like the Human Connectome Project, it makes its vast databases and search tools available to researchers around the world.

Now imagine hundreds, if not thousands, of other research groups contributing to our growing knowledge of the brain, and you will have some idea of the progress we are making. That progress is exciting and ongoing.

BRAIN/COMPUTER INTERFACE

Important progress is being made in brain studies and other technologies that will be needed to achieve Kurzweil's singularity. These efforts are attempts to solve medical, cognition or communication problems related to the brain, and are not specifically targeted toward achieving any kind of singularity. One of the key characteristics of the Kurzweil singularity is the brain/computer interface (BCI) using nanobots to download and upload information between the brain and computers.

There are many current approaches that allow direct communication between the human brain and a computer. Obviously, my brain is communicating now with my computer as I type these words. That doesn't count. We're talking about ways to connect without using my hands or voice – that is, just by thinking or even without conscious thinking.

This is what is generally meant by the term brain/computer interface.

The most non-invasive BCI technique today involves use of the EEG (electroencephalogram). With sensors attached to your scalp, an EEG can detect the electrical brain waves produced by brain activity. EEGs have been used for decades to detect points of origin of epileptic seizures, normal and abnormal sleep and waking patterns, and certain pathological brain conditions. These wave patterns have provided a great deal of information about normal and abnormal brains. Still, an EEG is too crude an instrument to tell what you are thinking or what you know. There have been EEG developments that allow a patient to control a wheelchair, a robot or a mechanical prosthetic device. These experiments require the patient to be trained to create specific EEG patterns by performing some mental task to represent specific commands. For example, a paralyzed patient can be trained to think about moving a computer cursor in a manner that causes a specific and detectable change in the EEG. A computer can then be programmed to move the cursor in response to detecting that EEG change.

Another example of using the EEG for BCI is a study reported in the journal *Science Reports* in which people were taught to generate two specific EEG patterns that could be communicated electronically to a computer. These signals, in turn, could non-invasively generate a magnetic impulse to the brain of a second individual. These people could thus cooperate in performing a simple task by communicating brain-to-brain through a computer interface. There is no direct application for this now, but it does demonstrate a wider potential for brain-to-brain communication through a computer.[112]

However, brain/computer interfaces using EEGs have not been deployed widely because of the non-specific nature of EEG signals. An extracranial EEG sensor (a metal electrode detector pasted to the scalp) is a blunt instrument because the detected brain waves represent the summation of millions, or even billions, of neurons. All that happens in the brain – from initiating movement and perceiving sensory input to memory, thought, consciousness, emotion and everything else – depends on the firing of neurons and how they are networked. To understand detailed brain activity, one must detect individual and small groups of neurons firing. This cannot be done with a typical EEG.

For that reason, there are two different approaches to going beyond current EEG technology to detect brain activity at a finer level of detail. The first is to improve the BCI resolution of scalp-based detectors using non-invasive techniques. DARPA is working on an initiative called the N^3 program (Next-generation Nonsurgical Neurotechnology) that utilizes a complex combination of light, ultrasound and other physical properties to both detect and communicate signals to the brain non-surgically, through the scalp.[113] The intent is to enable able-bodied military personnel to more rapidly communicate with computers in a variety of situations.[114]

The second approach is to bypass the blunting effects of the scalp, skull and other tissues and go directly to the brain using neurosurgery. The two most common methods require opening up the skull (craniotomy). The first one involves placing tiny needles into a portion of the brain to detect the activity of individual and/or small groups of neurons. The second, called ECoG (electrocorticography), involves placing sensors directly on the surface of the brain, either just over or just under one of the brain coverings (layered tissues called meninges). Both techniques are a huge improvement in precision mapping of brain activity compared to an EEG. Since direct needles and ECoG require opening the skull, they are much more limited in human studies.

Much of the current research with these techniques involves training amputees to use computer-controlled artificial limbs. The basis for these treatments is that muscle control of the limbs is generated in a specific part of the neocortex, called the motor cortex. All parts of the body are mapped into the motor cortex. ECoG or needle sensors can be placed to detect when motor cortex neurons for specific muscle groups are activated. These signals are then sent to operate the computer-controlled artificial limb. The results are encouraging, and this technology is increasingly used with both military and civilian amputees.

The somatosensory cortex is the equivalent to the motor cortex for physical body sensations. Experimentation on monkeys is now under way to place additional sensors on artificial limbs, which provide feedback to the somatosensory cortex to enable the amputees to *feel* using the artificial limbs.[115] This is important because virtually all movements depend on sensory feedback to enable precise movement. It is much easier to grasp something when you can feel what you are grasping, rather than doing it using only visual feedback. Even simply moving an object from point A to point B requires significant sensory input. The direction needs to be monitored with joint proprioception sensors, the speed of movement has to be governed, and the proper force needs to be applied depending on the object's weight and its resistance to movement. All these movements require coordination of thousands of motor and sensory neurons. Imagine the difficulty of trying to type something if your fingers were anesthetized. Current BCI technology cannot provide the normal richness of feedback to and from the brain. However, it can provide enough information to allow a person to learn how to control artificial limbs. Early experiments in humans to provide sensory feedback to amputees with BCI-controlled prostheses promise to greatly enhance that control.[116]

Equally exciting is the use of BCI in patients with spinal cord injuries. These individuals are either quadriplegic (all four limbs are paralyzed) or paraplegic (only the lower limbs are paralyzed). Their motor cortex still functions, even though there is no longer a direct nerve connection to the muscle groups. Signals from the motor cortex are sent to a computer that can control a prosthesis or a mechanical exoskeleton, or even sent to wires directly connected to the patient's muscle nerves, bypassing the spinal cord injury. With time and training, a patient can regain mobility by learning to control the system just by *thinking* of moving a limb. Although it is quite a bit more complicated than I've described here, these systems work. Patients think about walking and their legs walk. It's not as good as the body's original neurological system, but a lot better than being paralyzed. With time, these techniques will come increasingly close to reproducing normal movement and sensation.

Brain computer interface techniques are also allowing patients to communicate after losing their speech ability following a stroke or from diseases like ALS (Amyotrophic Lateral Sclerosis, or Lou Gehrig's Disease). These patients learn to control a computer cursor mentally that allows typing on a visual keyboard. Such experiments have shown that patients can generate messages at a reasonable typing rate of 25–30 characters per minute.

Even more exciting in speech BCI research are efforts to create synthesized speech directly from the areas of the motor cortex that control the muscles in the lips, tongue, larynx and jaw. These muscle movement signals are sent to AI software to generate the corresponding sounds. Although these efforts are still a long way from producing normal speech, these breakthroughs are at least producing intelligible speech.[117]

The next steps are to reduce or eliminate the need to open up the skull. Experimentation in mice is currently under way to use an injectable substance that forms what is called *neural lace* over the surface of the brain, eliminating the need for ECoG or needle sensors.[118] This neural lace is an electronic mesh placed directly on the neocortex surface or other brain areas for both recording and stimulation. It requires using a very narrow opening in the skull to inject the material. These efforts may be a step toward Kurzweil's singularity, but are still a long way from his vision of injectable, microscopic robots populating the brain.

Another type of BCI is the development of implantable devices for specific functions. Among the most successful of these are cochlear implants to enable some degree of hearing in the deaf. Hundreds of thousands of deaf patients have now learned how to use such implants to partially restore their hearing. These devices work by using an external sound receiver that is either worn behind the ear or attached to clothing. This device wirelessly transmits sound signals to a surgically implanted receiver next to the patient's auditory nerve in the cochlea (the part of the inner ear involved in hearing) in one or both ears. The implant translates these signals into electronic stimuli of the patient's auditory nerves in patterns that allow the brain to interpret the signals as speech or other sounds.

Another successful, but more complicated, sensory interface is an implant to restore vision in the blind. In this case, an external camera receives the incoming light information and wirelessly transmits the signals to a retinal implant. This implant contains an array of electrodes to stimulate the optic nerve with patterns that the brain can crudely interpret. This does not yet restore sight to anywhere near normal, but the technology will undoubtedly improve.

We talked in the previous chapter about the views of Elon Musk regarding the possible existential threat of AI. He becomes relevant again regarding brain/computer interfaces. In 2016, he announced that he was starting yet another company: Neuralink. Although his comments were somewhat secretive and vague, he clearly envisioned developing a non-invasive, high bandwidth, wireless BCI. At the time, he used the term *neural lace* and hinted at introducing it into the brain intravenously, thus invoking (intentionally or unintentionally) images of the Kurzweil vision. Nothing more was heard about the project until July 2019, when Musk and the Neuralink team came out of the shadows and publicly presented what they had been working on for the previous two years. They provided an update presentation in August 2020.

The presentations are available through Neuralink's website and are, in my opinion, amazing. Musk has assembled a first-class team of neuroscientists, physicians, nanotechnologists, software engineers and other technologists and has developed what appears to be the most advanced BCI technology to date. It has only been tested in animals – most recently pigs – as of this writing. In July 2020, the team received Breakthrough Device Designation[119] from the FDA, and its goal is to perform the first human clinical implementation in a quadriplegic patient in the near future.

First off, it is *not* the neural lace described earlier and it is *not* deployed intravenously. Instead, it is a vast improvement on the direct needle sensor technology. The needles are flexible polymer threads one-tenth the width of a human hair – roughly the width of a single neuron. They have engineered 1,024 of these threads into an integrated wafer that handles the neuron spike detection, software, communications, and all other functions necessary for wireless communication both to and from the sensors. These threads are barely visible to the human eye – too tiny and too delicate to insert manually – and require a newly invented robot to implant them into the brain. The robot is directed under human control to place each thread so as to avoid damaging any blood vessels. Even though the skull is immobilized during the process, the brain itself moves slightly with each breath and heartbeat, and so requires a very precise implementation.

The threads are attached to a coin-sized[120] circular device that is embedded in the skull and communicates wirelessly to a mobile device. The implantation occurs without general anesthesia and will be done as an outpatient procedure. The Neuralink team eventually plans to cover all sensorimotor functions, including speech generation, and then move on to other areas like the visual cortex and, ultimately, to deeper areas of the brain.

Although all of this technology is awaiting validation in human clinical trials, the progress already demonstrated in such a short time is truly remarkable. According to Musk, the two-way communication potential of his technology will ultimately enable uploading information back into the brain, just as Kurzweil predicts. He has provided no details regarding this possibility, so I will reserve judgment on its credibility. Although Musk has argued that AI is an existential threat to mankind, he now says that such a merger of computer information with the human brain could be a mitigation against that threat. He says it will allow us to "go along for the ride" as AI advances and leads to "AI symbiosis."

NANOBOTS

If and when the Neuralink BCI is deployed successfully, it still falls short of the Kurzweil vision. Kurzweil's singularity calls for nanobots that can be injected into the bloodstream and make their way into the brain. Then, somehow, they are able to detect the information stored in neurons, which can be wirelessly transmitted to a computer.

There is nothing even close to that being tested today. However, Kurzweil's 2005 vision has advanced considerably. In a major review article published in 2019, an updated description of how the Kurzweil singularity could be achieved is presented, based on the most recent advances in nanotechnology.[121] This revised description envisions multiple types of nanobots communicating with massive cloud computers. It is still largely speculation.

However, there is significant progress on nanobot use for medical purposes. One focus is to attach substances to specially prepared membranes of red blood cells or platelets, in a manner that shields them from the body's immune rejection system. These substances are of two general types. The first enables the direction or movement of these microscopic entities using sound waves or magnetic fields, creating what are called *micromotors*. The second type is drugs that could perform some therapeutic function when the micromotors come in contact with infectious agents or tumors. Another approach in experimental phases is to attach therapeutic substances to actual biological motors, such as sperm cells or benign bacteria, that move themselves with their natural, hair-like cilia. So far these forms of nanobots are being tested in laboratory settings and have not been deployed in clinical experiments in humans. However, it is only a matter of time before clinical trials will become feasible.[122]

One example in the pre-clinical phase is a promising experiment done at the Segal Cancer Center in Montreal. Researchers used what they called a 'bacteriobot' to treat a type of cancer in mice. This bacteriobot is a naturally occurring bacterium that swims in the bloodstream, propelled by a long hair called a flagellum. Anti-tumor drugs were attached to these bacteriobots, which were directed to the mouse tumor using magnetic fields. Once there, the bacteriobots were able to penetrate the tumor and deliver the drugs effectively.[123] The future of this type of nanobot cancer therapy is profoundly exciting. It has the potential to dramatically increase the effects of chemotherapy without the huge negative side effects of damage to healthy tissues. Similar experiments in mice have been used to genetically modify specific cells by delivering DNA-based substances rather than drugs – a form of nanobot genetic engineering. Experiments are also under way with nanobot anti-infectious agents that can target specific bacterial infections.

Perhaps the most amazing approximation of future nanobots was reported in January 2020 by a team of researchers at the University of Vermont and Tufts University.[124] Their creation, called xenobots, are microscopic robots made of living cells that are less than a millimeter in size. They are created to perform specific tasks, such as sweeping up small particles or carrying a small amount of a substance in a depression in their bodies. These xenobots are first created in a computer simulator, which is used to design their shapes and functions. The computer design is then translated into living microscopic robots by taking combinations of skin and heart cells from frog embryos and constructing a living version of the robots in the same shape and pattern as the computer simulation. This remarkable process is still in early *in vitro* studies, but its potential is limited only by the imagination.

These are just a few examples of the burgeoning field of medical nanotechnology, which has attracted its own research specialists, academic departments and dedicated journals. Compared to the Kurzweil vision, today's nanobots are primitive. None target the brain, none could cross the blood-brain barrier that normally prevents substances from entering brain tissue, and none could transmit or receive information. Nonetheless, the progress being made in medical nanotechnology is impressive.

THE SKEPTICS

Despite all the progress already made, Paul Allen and Mark Greaves, an eminent computer scientist, noted in their famous 2011 article, "The Singularity Isn't Near?"[125] that Kurzweil's predictions are based on the assumption that his Law of Accelerating Returns will enlarge our understanding of the human brain in the same way that Moore's Law predicted the growth in computing speed. Allen and Greaves state that applying this law to our understanding of the human brain is simply flawed. They insist that the brain is far too complex, and our grasp of it is far too primitive, to expect such rapid understanding. They describe a *complexity brake* that will prevent such growth. This is the notion that with such complex studies, the more you learn, the more you realize how much more there is to learn. In complex areas, breakthroughs are unpredictable and intermittent, and previous 'facts' get proven incorrect and need re-evaluation. Progress is spotty, bumpy, and certainly not on any exponential curve.

Two examples of such a complexity brake are in the study of quantum mechanics and astrophysics. Quantum mechanics remains a field with many mysteries and not on an exponential curve of understanding. Similarly, our knowledge of the astrophysics big picture – gravity, energy and matter – seems stalled despite our telescopes and space probes. Allen and Greaves conclude their article by saying, "By the end of the century, we believe, we will still be wondering if the singularity is near." Corey Pein, an investigative reporter, goes even further. He calls Kurzweil's 'singularitarianism' an intellectually fraudulent religion.[126]

Believing that the singularity is science is not the same as believing it will happen. The ASI-type singularity is the most frightening version. However, its likelihood is not high, and it's certainly not likely in this century.

The Kurzweil type singularity seems more a gradual improvement in BCI technologies and nanotechnology over many decades than some sudden, dramatic achievement. We're already downloading neuron spike patterns to a computer to represent intended movement. We're also uploading stimulation to the various sensory cortices to elicit touch, vision, auditory and other sensations. These are early first steps toward the Kurzweil vision.

But downloading thoughts, memories, knowledge, capabilities, emotions and consciousness is much less likely, although theoretically they should be patterns of neuronal spikes somewhere in the brain. That's where the connectome will come in handy. We know already from imaging studies and the vast databases of connectome studies that there is variation in the wiring of human brains – sometimes quite dramatic. Is that what accounts for the great differences between people in their cognitive capabilities? Those differences include mathematical ability, music skills, artistic creativity, motivations, personalities and all of the components of intelligence we discussed in the last chapter. Are all of these explained by their connectomes, or is there something else going on? We don't know.

The least probable capability in the Kurzweil vision will be uploading new knowledge or skills into the brain from a computer. That would require stimulating specific new networks of neurons and their synapses in predictable ways – something that would seem to have a tremendously high complexity brake.[127]

It is possible that, given enough time – perhaps a century or more – we will gain most of the understanding of complex brain functions like consciousness, memory, thinking and creativity. After all, the brain consists of anatomy, chemistry and physiology like every other organ, except orders of magnitude more complex. As the astronomer and pop-science icon Carl Sagan stated in his book,

Dragons of Eden, "My fundamental premise about the brain is that its workings - what we sometimes call the 'mind' - are a consequence of anatomy and physiology, and nothing more."

Not everyone agrees with this. The chimpanzee will never understand prime numbers or nuclear fusion or the Pythagorean Theorem, no matter how long we let evolution do its thing. The chimp brain simply isn't capable, and the odds of it undergoing the same mutations and natural selection humans experienced seems quite remote. Likewise, perhaps there are things that the *Homo sapiens* brain simply will never understand, like the origin of the universe, the outer boundary of the universe or quantum mechanics. There is a group of people called *Mysterians*, the most notable of whom is Colin McGinn,[128] who would put understanding of the human brain, particularly consciousness, in the unknowable category. To Mysterians, this is not a complexity brake. It is a definitive complexity barrier.

Nor can we count on Darwinian evolution to improve our ability to understand the brain. Darwinian evolution is not about increasing intelligence; it is about increasing the ability to procreate. For a large part of our evolution, those two were synergistic. That is no longer the case. There is no evidence that the most intelligent of our species, no matter how measured, are the most prolific procreators. Natural selection does not appear to be selecting for intelligence any longer. Perhaps our ultimate irony is that the human brain will never understand the human brain.

The singularity is not near.

THE SINGULARITY AND THE FOURTH GREAT TRANSFORMATION

Even if the singularity is achieved one day, could it be considered the *Fourth Great Transformation*? That is, is an ASI or a Kurzweil-type cyborg a new species of human? With regard to an ASI computer, it could not be considered *Homo nouveau*. First of all, it would not be human – at least by any reasonable definition of the *Homo* genus. Second, it would not even meet anyone's definition of life (even though it might meet its own definition of life). But what about the Kurzweil cyborg version of the singularity? Would that be a new human species? The answer is no. There is nothing about the Kurzweil cyborg that would necessarily meet any definition of a new human species. Those individuals would still interbreed normally with any other humans. There would be no reproduction isolation barrier, as described in Chapter 1. They would not be evolving independently.

One of the most enjoyable aspects of Kurzweil's book, *The Singularity is Near*, is his fanciful dialogues at the end of each chapter. He creates an imaginary conversation between himself and people like Charles Darwin or Sigmund Freud, or just some future person, to illustrate the chapter's key points. So, in the spirit of true admiration, I would like to emulate Kurzweil's brain with an imaginary conversation of my own.

Ray (Kurzweil):
So, Don, now that you've reviewed the literature on neuroscience, what do you think about my prediction?

Don:
Which one? You've made so many.

Ray:

About reaching the singularity by 2045! Isn't that what this chapter is all about?

Don:

Oh, of course – that prediction. Well, the progress is amazing! We have learned an incredible amount in the past 15 years, since you published your book. We know so much more about the human brain connectome than we did back then, which seems like the dark ages now. Numerous brain functions are being simulated in a computer. We're doing amazing things with the brain/computer interface, like controlling artificial limbs and interpreting speech from the electronic signals from brain implants. Nanotechnology is reaching a point where real animal experiments to solve real clinical problems are occurring in genetic engineering, oncology and other areas, using what you call nanobots. You're a true visionary.

Ray:

Thanks, but you didn't answer my question. How close have we gotten to the singularity in these 15 years? And is that close enough to reach it by 2045?

Don:

That's a tough question. It's really difficult to put a number on that.

Ray:

I know it's tough, but you're the one writing this book. Take your best shot. Make a reasonable guess.

Don:

In spite of all this great progress, I still don't see how we bridge the gap between connectomes and everything else we're learning to understand – reflective consciousness, pattern recognition, fictive thinking, common sense, and all the other things we need to solve to exceed human intelligence. And, I still don't get how injecting nanobots into the blood stream will lead to downloading my brain into a computer ... and, even more unlikely, uploading new knowledge into it. That one sounds a lot like Douglas Hofstadter's critique that I quoted at the beginning of this chapter. If you're going to force me to put a number on it, I'd say we're only at about 1% of the way there at best. At this rate, it will take 1,485 years to get there (15 times 99).

Ray:

I thought you said you read my book.

Don:

I did. It is a real inspiration to me.

Ray:

Well, you obviously missed the point about the Law of Accelerating Returns. You're still stuck in the human intelligence flaw of linear thinking. The law of accelerating returns says that technological progress is not linear, but exponential. And neuroscience technology is not likely to be any different than any of the other technology examples I gave you in my book. Progress toward the singularity will double every two years. If you're correct that we have gotten to 1% in only 15 years, then we'll get to 100% in just 14 more years, or 2034. That's 11 years earlier than even I predicted. I'd hate to think what Douglas Hofstadter would say about you!

CHAPTER 6

GENETIC ENGINEERING: THE SECOND ULTIMATE TOOL

"I think it is no exaggeration to say we are on the cusp of the further perfection of extreme evil, an evil whose possibility spreads well beyond that which weapons of mass destruction bequeathed to the nation-states, on to a surprising and terrible empowerment of extreme individuals."[129]

—**BILL JOY**, co-founder of Sun Microsystems, speaking of genetic engineering, nanotechnology and AI in *Wired* magazine

Genetic engineering will be the tool that we use to create *Homo nouveau*. AI will be the enabler of that tool. To understand these two points, it is necessary to understand the complexities of genetic engineering and the role AI will play in resolving them.

Evolution happens naturally, as random mutations and other changes occur in the genome of a species. These changes alter traits or characteristics that enable the species to better adapt and make it more likely to reproduce. That is natural selection. In contrast, genetic engineering is a process in which *Homo sapiens* – and only *Homo sapiens* – deliberately alters the genome of a species to produce a desired trait or characteristic. The goal of genetic engineering is not necessarily to improve procreation but rather to provide some benefit to us. Genetic engineering is not natural. It is artificial. It is not meant to be random. It is deliberate.

A QUICK REVIEW OF THE BASICS OF GENETICS

For those of you who are comfortable with the basics of genetics, as covered in Chapter 1, you can skip to the next section. I will provide a quick summary here.

- The characteristics of all plants and animals are determined by their genes.
- Almost all of an organism's genes are contained in the nuclei of all cells, in structures called chromosomes. Humans have 23 pairs of chromosomes, for a total of 46. A small number of genes are also contained in the mitochondria of cells, which are outside of the nucleus in the body of each cell.
- Genes consist of DNA. DNA consists of four nucleotides, labeled A, T, C and G. The total DNA of an organism is called its genome.
- The most important function of genes is to produce proteins. These are called coding genes. Proteins perform all of the most important functions in an organism, including structures, organs, enzymes for metabolism, hormones and virtually everything that is needed for living.
- Proteins consist of 20 amino acids. There is a three-nucleotide code called the codon that is used to translate nucleotide sequences into amino acids.
- To create a protein, the DNA of genes is first translated into RNA in the nucleus, which then enters the body fluid of the cell to get translated into proteins using the codon.
- In addition to the coding genes, which constitute less that 2% of our genome, there are other types of genes that produce substances that control the expression of each gene. These are sometimes referred to as non-coding genes. Expression determines at what stage in the life cycle of each organism, and in which tissues,

that gene is functioning. This expression-controlling part of the genome is called the epigenome. Together, the coding genes and epigenome constitute less than half of the total genome. The remainder is sometimes called *junk*, although we are continually discovering important functions or impacts of components of that junk.
- All plants and animals use the same four nucleotides in DNA and the same 20 amino acids in proteins. The only difference between them is the sequence of the nucleotides in the genome.
- In sexually reproducing plants and animals, each cell contains two copies of each gene, one inherited from the mother and one from the father. Mitochondrial genes are inherited only from the mother.
- Inheritance of traits is through Mendelian genetics, according to dominant and recessive rules of inheritance, as described in Chapter 1. The traits are called the *phenotype* and the composition of the genes is called the *genotype*.
- Genetic engineering is a tool created by *Homo sapiens* to alter the genes of any organism, including humans.

THE FIRST GENETIC ENGINEERS

We first attempted genetic engineering long before we understood anything about genes, inheritance or evolution. It began about 12,000 years ago, when we transitioned from hunters and gatherers to farmers. We started to selectively breed plants and animals to make improved food products, stronger animal work tools or better traits in something else. We learned to crossbreed a plant or animal having a desired characteristic with another with the same characteristic to most likely produce an offspring with the same characteristic. Not always, but most likely. So, if you want a fast horse, find a fast stallion and a fast mare and breed them. Likewise, if you want a red tomato, pollinate the reddest tomatoes with pollen from other very red tomatoes.

That's how we changed wild grasses into the wheat, barley, rye and rice we have today. And how we got our beautiful, large-grained yellow corn from an early wild maize with tiny kernels called teosinte. And how we got our beefy meat cows, large-udder milk cows, and chickens that produce an egg a day, rather than the original one egg a month. The human diet has been transformed over millennia by this form of genetic engineering.

By selectively breeding dogs we've gotten today's dachshunds, labradoodles and westies. Likewise, the majestic Budweiser Clydesdales are the result of generations of horse breeding. Some of our most beautiful flower varieties are the result of selective breeding. Countless other plants and animals have been so altered just to please *Homo sapiens*. This is sometimes referred to as *artificial selection*, where the selection process is done by humans and not the environment. Even so, those desired traits still had to first occur in nature before being selected.

But what if you want something that doesn't exist in nature to begin with? Suppose you want to develop a breed of dog that weighs ten pounds from one that normally weighs twenty pounds? As discussed in Chapter 1, the French biologist Jean-Baptiste Lamarck thought this could be done by deliberately under-nourishing dogs and then interbreeding them. He believed that traits so acquired during an animal's lifetime were passed on to their offspring. Lamarckian evolution was mostly debunked in the 19th and 20th centuries, when Mendelian inheritance through genes became well established as science. However, we now know that starvation can alter the epigenome in ways that could be passed on to future generations. But that would not be the most efficient (or humane) way to genetically alter a species. Instead, to get ten-pound dogs, the naturally smallest dogs (the 'runts' of each litter) would need to be selectively bred over many generations to eventually evolve a miniature version of that breed.

Darwinian evolution and natural selection are random processes whose only goal is to increase the number of offspring. It takes many generations, and sometimes thousands or even millions of years. Artificial selection has the goal of bringing out some particular trait desired by *Homo sapiens* and occurs much faster, although it still may require many generations. But, more importantly, it requires the random process of natural mutations to produce the desired trait. Being humans, however, we wanted more control to eliminate the random part and speed up the process.

MODERN GENETIC ENGINEERING

During the first half of the 20th century, we gained a deeper understanding of how genetic changes occur. As covered in Chapter 1, we learned that inheritance is controlled by substances in our chromosomes called genes. Those genes consist of DNA, and that DNA consists of nucleotides. Most importantly, we discovered that a change in those nucleotides caused either a positive, a negative or no effect (in most cases) on the organism.

So, if we could determine which nucleotide sequence caused which trait, then we would need only to change that sequence to create or improve the trait. Like typing on a typewriter to write this book. Only, it didn't turn out that way. We didn't know how little we really knew about the genome and how difficult changing it would be.

In the 1960s, our initial attempts to go beyond artificial selection were crude. We irradiated plants, insects and other animals to produce random DNA mutations. Then we observed the results for effects. We wanted to learn more about how genes control traits, but also how to produce something useful. Sometimes we succeeded, but this was still a random process that just sped up the rate of mutations. We weren't controlling anything.

In the 1970s, experiments with bacteria showed that a DNA segment containing a gene from one bacterium could be patched into the DNA of another bacterium of that species to form what is called *recombinant DNA*. The gene would function in the new bacterium just as in the original one. For example, resistance to a specific antibiotic could be transferred this way between bacteria. Further experiments showed that recombinant DNA even works between two different bacterial species. Finally, it was determined that recombinant DNA can pass virtually *any* gene from any plant or animal to any other plant or animal. As noted in Chapter 1, this is because all plants, animals, bacteria and other life forms are alike at the molecular level, except for their DNA sequence. The modern era of genetic engineering was born.

One of its firsts benefits was to improve the treatment of diabetes. Insulin is a protein produced in the pancreas under the control of a gene. It's a key player in our body's metabolism of carbohydrates, fats and protein, and it controls the amount of sugar in the blood. Diabetes is a disease that hinders the body's ability to produce or respond to insulin. Some diabetics had to be treated with insulin laboriously extracted from a pig or cow pancreas. This meant that 56 million pigs and cows had to be slaughtered annually to meet the demand. This was expensive and problematic, as animal insulin sometimes caused allergic reactions in humans because it is a slightly different protein.

In 1982, the biotech company Genentech engineered bacteria to produce human insulin, the first of many pharmacologic uses for this technology. In a laboratory, Genentech manufactured the DNA sequence that produces human insulin and patched it into a common bacterium called *E. coli.* Within days they had vats of *E. coli* producing human insulin cheaply and in the exact form that is safe for humans, without adverse allergic effects.[130]

In 1994, the FDA approved the first genetically engineered food. The Flavr Savr tomato was genetically engineered using a bacterial gene that delayed spoiling. Although the Flavr Savr tomato was not a commercial success, it opened the door to thousands of other genetically engineered plant foods that are on the market today. These foods provide improved traits such as resistance to disease, tolerance of herbicides, improved appearance, better taste, increased nutritional value, drought resistance and many others.

The basic idea is to identify a trait anywhere in nature that might be useful in another organism, isolate the gene that produces that trait, and then splice it into the DNA of that organism to transfer the trait.[131] For example, transferring the gene for scorpion venom into cabbage makes that cabbage resistant to being eaten by caterpillars, yet is not toxic to humans. We took genes from monarch butterflies that like milkweed and put them in flies that usually die from eating milkweed, so now these flies can eat milkweed safely.[132] We have genetically engineered trees to remove certain groundwater contaminants, bananas that produce vaccines, cows that produce less methane flatulence, pigs with eco-friendlier poop, goats that can produce spider silk in their milk and other goats whose milk contains specific pharmaceuticals. These last are called *pharm animals.* There are *pharm plants* as well. The potential of genetic engineering is limited only by the imagination and the availability of the desired gene somewhere in nature.

However, it took much longer to certify a genetically engineered animal for food. In 2015, the FDA approved a genetically engineered salmon with a gene that produces higher quantities of growth hormone. This causes the salmon to grow twice its normal size while using far less feed than natural farm-raised salmon. The FDA declared the fish safe to eat, but it was prohibited from U.S. sales because of bureaucratic delays regarding GMO labeling. AquaBounty, the producer of the salmon, states that its product will be labeled as genetically engineered regardless of government labeling requirements. It is already being sold in small quantities in Canada and is expected to reach American stores in 2021. It is unclear if U.S. groceries and, more importantly, American consumers, will accept it. It has taken the company more than 30 years to reach this point since it first developed the genetic modification.

Public resistance to genetically engineered foods – sometimes referred to by critics as 'Frankenfoods' – remains strong, even though multiple studies have determined there is no evidence of harm from eating them. That holds true for both plant and animal GMOs. On the other hand, there is evidence of harm from *not* accepting such foods. A good example is the struggle to get approval for growing Golden Rice in developing countries. More than 20 years ago, a variety of white rice was genetically engineered to be rich in vitamin A, but this so-called Golden Rice has been blocked by various activist organizations that oppose GMOs. As the science writer Ed Regis told *The Guardian*, "Had it been allowed to grow in these nations, millions of lives would not have been lost to malnutrition, and millions of children would not have gone blind."[133]

In 1999, environmental activist Mark Lynas and a group of like-minded friends took their machetes in the dark of night and hacked down a field of experimental genetically modified maize in England. He was one of the leaders of extreme anti-GMO radicals trying to save the world from this perceived existential threat to humanity. Twenty years later he published *Seeds of Science ... Why We Got It So Wrong on GMOs*. In this well-researched book, he details not only his own conversion on this issue, but the massive damage done by the anti-GMO resistance, especially in Africa. He describes the great difficulty, which still continues, involved in getting approval to plant genetically modified drought-resistant and pest-resistant food plants. That resulted in increased use of pesticides and contributed to widespread malnutrition. According to Lynas, the irony is that this resistance is from well-meaning charitable and non-government organizations purporting to promote organic foods and healthy lifestyles. Their tactics, however, have been riddled with outrageous and unscientific claims regarding the dangers of GMOs.

The debate is further complicated by the various international definitions of a GMO. What if the organism's genes are modified by using radiation or another non-genetic-engineering mechanism? What if the modified genes are from the same species? Or, from a related species, as is the case with the AquaBounty salmon, whose modified genes come from other fish species? The subject is quite nuanced and not well understood by the general public, including those for and against GMOs. Recent research has shown that we can now genetically engineer the exact same changes in a plant by altering *how* the proteins are produced by a gene rather than changing the gene itself.[134] That will result in genetically modified organisms that don't meet any current GMO definition. It is not clear if altering only proteins and not DNA will resolve this debate.

The resistance to eating genetically engineered animals is stronger than the resistance to eating genetically engineered plants. There is no rational basis for this. Note that there is no public resistance to genetically altered animals created by selective breeding rather than genetic engineering.

Still, all this does not mean that GMOs pose no risks. That will probably never be possible. Genetic engineering is too complex, and we don't yet know everything we need to know about it. For example, a company in Minnesota genetically engineered a breed of dairy cows to have no horns, which would decrease injuries to farmers. The process used a bacterial-based vector (see below for a discussion of vectors) to deliver the hornless gene. It was later found that some of these hornless cows also inadvertently carried a gene from the vector that provided antibiotic resistance in the bacteria. Somehow that gene was entered into the cow's DNA along with the hornless gene. The presence of this unexpected gene in the cow would not likely harm a human if they ate its meat (which was never the intent) or drank its milk. It does illustrate, however, that unexpected alterations in DNA are possible with genetic engineering. The genetic engineering tool used in this case was one of the older ones, and newer tools have less risk of this problem. Nonetheless, I cannot unequivocally state that there is no risk. The same can be said for virtually every medical procedure done on humans today.

Despite persistent resistance, GMO-based food is now an established component of the human diet throughout the world, as are GMO-based pharmaceuticals such as insulin. To get to the *Fourth Great Transformation*, however, we will need to cross the Rubicon to the ultimate stage of genetic engineering. That is to create a special GMO called a GMH – the genetically modified human. This is where the complexity, and risk, escalates exponentially.

THE FIRST GENETICALLY MODIFIED HUMANS

OTCD (ornithine transcarbamylase deficiency) is a rare disease you've probably never heard of. Many children born with the severe form of it die within five years. Others can be maintained with intense dietary, medication and sometimes dialysis therapy. It is caused by a mutation of a single gene that makes the liver unable to produce an enzyme to metabolize ammonia. The ammonia build-up kills the patients.

In 1999, the University of Pennsylvania was a pioneer in the study of genetic disorders and the design of early experiments to attempt genetic engineering cures. OTCD was an ideal candidate for such a cure, since the genetic error in a single gene was known. It was thought that delivering a normal gene copy to the liver cells would cure the disease, because they would then be able to produce the normal enzyme. The best way to deliver these genes was to attach them to viruses that had been altered so they could no longer cause a viral illness. These viruses still retain their natural ability to enter human cells and integrate their DNA to produce proteins. Such delivery-vehicle viruses are called *vectors* in the treatment of genetic disorders. Later, we'll discuss other types of vectors that can also deliver recombinant DNA to targeted cells.

In the proposed OTCD experiment, a virus was modified for that purpose that contained the normal OTCD gene. These modified viruses would be injected directly into the artery feeding the patient's liver. Animal experiments had already been safely performed using this protocol.

The next question was which patients to recruit for the initial experiments. The logical choice was infants with the full-blown disease who were destined to die in infancy. However, the university ethicist argued that it would be unfair to ask parents of such children to volunteer. It was felt these desperate parents wouldn't be able to make an informed decision about the risks.

Jesse Gelsinger was an 18-year old with a mild form of the disease. It was mild because only some of his liver cells had the deficiency and others were able to produce the normal enzyme – a condition called mosaicism. With the proper diet and daily medication, he was able to live a normal life.

So, it was decided to recruit patients like Gelsinger, who did not need a cure and would be more likely to provide a reasoned informed consent. The study was approved by all of the appropriate review bodies.

Gelsinger volunteered and was the 17th patient entered into the study. No serious problems had occurred in the previous participants. However, Gelsinger experienced a massive immune reaction to the virus and died within days of treatment. This death was the first reported in a human genetic engineering experiment. It shocked the world and set the industry back in ways that are still being felt today. For one, there were no further attempts to genetically treat OTCD patients until very recently.

In retrospect, Gelsinger was probably not the first death in a human genetic engineering experiment. There had been previous experiments for other genetic disorders and patients had died in some of them. Those deaths were usually attributed to the underlying disease rather than the treatment itself, and were just assumed to be treatment failures. However, distinguishing a treatment failure from a treatment-caused death is not always easy in very sick patients.

Not all of these early experiments were complete failures. The first human gene therapy was on a patient with severe combined immunodeficiency syndrome, another rare genetic disorder. This disease is caused by a gene mutation that fails to produce an important enzyme in white blood cells for immunity. Lack of this enzyme results in increased infections, ultimately killing the patient.

In 1990, researchers at the National Institutes of Health extracted white blood cells from one of these patients and, using a virus vector, introduced a normal gene copy into those cells in the laboratory.[135] Then they injected the cells back into the patient. The patient did experience fewer infections following the treatment, but the success was short-lived. White blood cells are produced in the bone marrow from hemopoietic (blood) stem cells. The genetically altered white blood cells died out naturally and were replaced by abnormal cells still being produced by the patient's abnormal stem cells. The patient lived a normal life afterward but had to undergo repeated therapies. So, this could only be called a treatment and not a cure. To fix the condition permanently, recombinant DNA would have to be delivered to the bone marrow stem cells – a more difficult and risky procedure.

This lesson was applied to a different genetic immunological disorder that also involved white blood cells. This time, stem cells from ten patients' bone marrows were removed, genetically altered to contain a normal copy of the gene, and then the altered cells were infused back into the bone marrow. Nine out of the ten patients were completely cured. Their genetically altered blood stem cells populated the bone marrow and continually produced normal white cells. Unfortunately, three years after the experiment, two of the successfully treated patients developed leukemia. Further study determined that the viral vector that introduced the normal gene into stem cells not only changed the target gene, but also unexpectedly altered another gene that led to the leukemia.[136] This phenomenon is called an *off-target mutation*, and is a major problem still plaguing human genetic engineering.

These early problems caused a dramatic slowdown in human clinical trials. Although encouraging in some ways, the catastrophic consequences overwhelmed opinions in both the scientific community and the general public. Clearly, genetic engineering wasn't going to be as easy as first thought.

THE CHALLENGE

To genetically engineer a cure to a disease, the exact genetic abnormality must first be identified. To that end, there has been great progress in sequencing the genome quickly and cheaply. The first complete human genome was sequenced in 2003 as a result of the Human Genome Project. Costs and speed have improved exponentially, to where we now have databases of millions of human genomes throughout the world. A similarly exponential increase in data processing power has also enabled identification of more than 10,000 genetic disorders caused by mutations in a single gene.[137] Certainly, we have our work cut out for us.

Today's human genetic engineering is targeting only disorders caused by a mutation in a single gene, and only involves a small fraction of those 10,000-plus known disorders. Also, we are only attempting cures affecting the somatic cells of a single individual rather than germ cells, which would be passed on to children.

If the gene is autosomal recessive (i.e., not sex-linked – see below) and the patient has inherited the abnormal gene from only one parent, there is no need for genetic engineering. This is called the *carrier state* and the patient does not have the disease, since the patient's normal copy of the gene produces the desired protein. Only patients who have inherited the abnormal gene from both parents would be candidates for genetic engineering, since they completely lack the needed protein. Sickle cell anemia and cystic fibrosis are examples of these ailments. The goal here would be to either insert a correction to one gene copy or replace one copy entirely with a normal copy (called a knock-in), so that the normal protein will be produced.

If the gene is dominant, though, the patient only needs one copy to cause the disease, since the abnormal gene is producing a harmful protein. Huntington's disease is an example of this, and the treatment goal is different. The abnormal gene must be corrected or somehow blocked (called a knockout) to prevent the creation of the harmful protein.

Humans have 23 pairs of chromosomes in every cell. One of those pairs is called *the sex chromosomes* since it determines the sex of the individual. Males inherit an X chromosome from the mother and a Y chromosome from the father. Females inherit an X chromosome from both the mother and the father. There are many genetic disorders that are called sex-linked since the abnormal gene occurs only on the X chromosome. This creates a special situation with dominant and recessive genes. Most sex-linked genetic disorders are recessive. If the abnormal gene occurs in a female, since she has two copies of the X chromosome, she must inherit the abnormal gene from both parents to have the disease. This is rare. However, if it occurs in a male, since the male has only one copy of the X chromosome, he will have the disease. Genetic engineering would then be needed to correct or replace that abnormal gene. The various forms of hemophilia are examples of this type of genetic disorder, which explains why hemophilia is much more common in males than females.

The first challenge of genetic engineering is to determine the abnormality causing the disease and then select the proper treatment protocol. Remember that every cell in the body has copies of every gene, but because of the epigenome, only a small fraction of the total genes is expressed (operational) in any given tissue. That is, brain cells only use genes that are needed in a brain, lung cells only use genes needed in a lung, and so forth. The second challenge, therefore, is to target the genetic engineering to the tissue where the abnormal gene is expressed. For example, if the abnormality is in white blood cells, the treatment needs to target the bone marrow where the white blood cells are produced. If the abnormality is in an enzyme produced by the pancreas, the targeted cells need to be in the pancreas. That is why the OTCD study injected the virus with the recombinant DNA into the hepatic artery, which feeds the liver. The gene related to OTCD is expressed primarily in liver cells.

VECTORS

It is not enough to just understand a genetic defect, how it causes disease and where in the body it is active. We must also find a way to fix the gene in that location. Whether we wish to add a normal gene (a knock-in) or remove an abnormal one (a knockout), we must direct the treatment to where that gene is expressed. This remains one of our biggest challenges. The solution so far has been to use delivery agents called *vectors*.

We already covered the viral vector used in the OTCD experiment – the one that led to the death of Jesse Gelsinger. Viruses are ideal vectors since they can enter virtually any cell in the body and, depending on the virus, can target specific cell types, like those that comprise the lung or the gastrointestinal tract. Remember, viruses only survive because of their ability to enter the cells of living organisms (called hosts) and take over the ability of those cells to produce proteins from their own DNA or RNA. If that DNA or RNA has been genetically engineered to contain a normal human gene, then that gene will become active in the host human cell.

Since viruses also cause disease, their disease-causing ability must first be rendered inactive, yet still retaining their capability to enter cells. Or benign non-disease-causing viruses can be used. In either case, viruses are the most widely used vectors to deliver the corrected DNA sequences. As we have already learned, this is a difficult and complicated process that involves risks. One risk is that the patient may be allergic to the virus itself. Another risk is the possibility of off-target DNA alterations, which could cause many different types of problems, including cancer.

Even if an appropriate viral vector is used, it still must be delivered to the tissue expressing the abnormal gene. That could be by an intravenous injection, which eventually circulates to every tissue, or a more targeted injection into an artery serving the specific organ. The vector could also be injected directly into the appropriate tissue. The first-ever commercially available genetic engineering-based drug is Gendicine. It was developed and approved in China in 2003 for the treatment of head and neck cancers. It used a viral vector and is delivered either intravenously, into an artery, or by direct injection into the tumor. The first FDA-approved therapy to be directly administered to a patient is called Luxturna. It was approved in December 2017. This involves direct injection into the retina of a viral vector containing the corrective DNA to correct for a mutation causing blindness.[138]

A promising variation has emerged in recent years. Why not use the body's own immune system to target genetic disorders? As mentioned earlier, cancer results from the accumulation of dozens of mutations in normal cells. When the right number and type of mutations accumulate, the cells escape their normal controls and multiply rapidly. They invade surrounding tissue and then metastasize to distant areas. Each cancer has its own unique genetic makeup, which is abnormal because of those specific mutations.

The new vector is the body's T-cells – white blood cells that normally circulate and seek out infections. These T-cells contain DNA information about an infectious agent and seek a match. A new approach to cancer is called CAR-T (chimeric antigen receptor T-cells). A patient's T-cells are extracted from the blood and then genetically engineered with a portion of a cancer's DNA. A viral vector is used to introduce the cancer target code into the T-cells. These altered T-cells are incubated to produce billions of copies, which are then injected back into the patient. They circulate and seek out the matching genetic code of the tumor and destroy the cancer cells.

The early results are promising, although not without complications. To date, there have been hundreds of clinical trials using CAR-T, mostly targeting various blood cancers. There are typically cancer remissions after initial treatment, but the cancers often return later because they've mutated to escape the T-cells' targeting. There are also toxic side effects that vary from mild to life-threatening. Since many patients have already undergone chemotherapy, they may not have sufficient T-cells remaining to be genetically engineered. Finally, it takes weeks to prepare the cells for re-injection, which is an unacceptable delay for some patients.[139]

For these reasons, efforts are under way to develop an alternative type of CAR-T in which any person can donate T-cells, which can be genetically engineered and re-injected into any patient. If this proves feasible, it would allow storage and delivery of donated T-cells similar to the way blood banks operate.

The first FDA-approved CAR-T therapy is a drug called Kymriah, for the treatment of a certain type of white blood cell cancer. Although clinically effective, its high cost ($475,000 per patient) and small number of eligible patients are limiting its commercial success. Another CAR-T drug,

Yescarta, has also been approved to target related malignancies. It's also expensive, and it is still too early to gauge its long-term success. Other CAR-T cancer treatments have been approved and many more are in the R&D pipeline.[140]

Most of this CAR-T immunotherapy is directed at cancers of the blood and bone marrow. Efforts are now under way to modify CAR-T therapy to target solid tumors, like breast and prostate cancers. This is a more difficult problem, but early experiments are progressing.

Vector technology will certainly improve for cancer and other types of genetic engineering. One interesting approach being studied in animal models and *in vitro* human cell models is the development of human artificial chromosomes (HAC). What if we could create a 47th chromosome to contain just the new genes that we want to introduce?

That is exactly what the HAC experiments are now doing. There are some potentially huge benefits if we can pull this off. It gets rid of those pesky virus vectors and also allows us to incorporate entire genes rather than just the snippets we use now. The benefit of using large genetic segments is that, in addition to the gene itself, all of the epigenetic code that regulates gene expression would be included. The HAC could be installed in any convenient tissue, like bone marrow, to avoid the need to deliver the new genetic code into tissues that are difficult to access. Without going into detail, we are still a long way from allowing actual clinical trials in humans.

The ultimate vector could be some form of nanobot (see Chapter 5). These would be microscopic robots injected into the bloodstream that are programmed to seek out the exact cells targeted for genetic engineering, enter those cells and deliver the genetic fix. At present, there is no reason to expect this becoming reality in any predictable timeframe.

MODERN EUGENICS?

Obviously, we want to find cures for people born with serious genetic diseases. Although there are treatments involving diet or drugs that address the symptoms of many of these disorders, they are not cures. Correcting the genetic defect is the cure, and this has motivated virtually all of the approved clinical trials to date. Further, the focus has been on genetic disorders that are fully understood and involve a mutation in only a single gene. The clinical manifestations of these 10,000-plus single-gene genetic abnormalities vary from mild inconveniences to physical disabilities to shortened lives. Factors that determine which diseases to tackle first are its seriousness, its prevalence and the ability to target the tissues involved.

Diseases like diabetes, Alzheimer's and heart disease have clear genetic influences, but are not caused by a single gene mutation. Multiple genes are known to have a statistical impact on these diseases' likelihood, but there are unknown genetic and other factors that also contribute. At this time there are no clinical trials aimed at these multifactorial disorders. The complexities of genetically engineering these disorders are too great given our current capabilities.

Genes do not just impact diseases; they control everything about you. They affect your appearance, your physical and mental abilities, and virtually everything else that makes you the person you are. Genetic engineering could also change those things. Theoretically, it could change your eye color, skin tone, athletic ability and even intelligence (whatever that is). These changes would be called *human enhancement* rather than disorder treatment.

There is near-universal agreement against using genetic engineering for human enhancement. Today's genetic engineering is not without risk. There is no reason to risk off-target cancer-causing mutations or other complications for those 'benefits.' That calculation is different for disease-causing mutations. We may never reach 100% safety, but the risk/reward calculation for curing a genetic or life-threatening disorder justifies a certain amount of risk. Besides, most human enhancement genetics are in the multi-gene category, and those treatments are still a long way off. We do have the capability to alter multiple genes, but we're not ready to do that to humans for any reason. Finally, there are serious ethical issues associated with human enhancement that bring up all the negatives of earlier eugenics movements, which had significant racial discrimination components. We aren't likely to go there anytime soon.

Or are we?

PRE-IMPLANTATION GENETIC DIAGNOSIS

Pre-implantation genetic diagnosis, or PGD, is not genetic engineering. Yet, it is changing our thinking about many related issues. PGD is used in conjunction with *in vitro* fertilization (IVF). IVF, now a routine process in most countries, enables couples with infertility problems to have children. Tens of thousands of children are born annually using IVF.

In IVF, eggs from a female are removed from her ovaries and, in a laboratory setting, are combined with sperm from a male to create fertilized eggs called zygotes. The zygotes are incubated for 2–6 days and then implanted into a uterus to develop into a full-term newborn. PGD occurs during this incubation period, after fertilization but before implantation.

During this period, the zygote starts as a single cell and begins to divide into a multi-cellular human embryo. At this early stage, a cell can be removed to perform a complete genome analysis. This does not harm the zygote's development, and does provide complete genetic information about each zygote to determine which one or ones to implant.

The obvious use of this information is to screen for serious genetic disorders. No argument there. But what about looking at other characteristics in the zygotes? For example, sex. Although it would be sexist to describe gender selection as human enhancement, depending on which sex one desired, it is already being done in IVF clinics. What about choosing for eye color? That seems closer to human enhancement. That is also being done in some IVF clinics. Are we already on the slippery slope toward accepting human enhancement? What if we start selecting zygotes with known genetic determinants of tall stature, increased athletic ability, longevity, higher intelligence or some component of it like math ability?[141] Unlike genetic engineering, this has no risk of off-target mutation or other problems. If we allow it in PGD, why wouldn't we allow it in genetic engineering once we get good at it?

Human enhancement criteria are already being used to select embryos during PGD. There are companies that provide statistical analyses during PGD for complex genetic diseases as well as non-disease traits, such as height. Although they only provide probabilities and not exact predictions, these analyses are already being used to select which embryos to implant. These companies will get better at their predictions. Once genetic engineering of embryos is demonstrated to be safe, there likely will be little resistance for its use for human enhancement. The only questions are when, what, why and in what countries.

THE DISCOVERY AND USE OF CRISPR

So far, our discussion about manipulating human genes, knocking out bad genes and patching in normal genes has been superficial regarding the actual technology. Understanding those complexities will not be required to understand the *Fourth Great Transformation*. Still, there are some basic facts that will help with understanding this conversation and the terminology now appearing in the lay media. That is important because the public will ultimately be asked to weigh in on key ethical, political, financial and scientific issues being raised by genetic engineering.

No two humans have the same genome – not even identical twins. Although we all have the same genes, there are many differences among us in the DNA sequences of those genes. That's why we have variations in eye color, height, weight, skin color, facial appearance, athletic ability, mental characteristics and so on. If and when we ever get to genetically engineering for human enhancement, those normal genes would all become targets.

Genetic mutations can cause any characteristic to be out of normal range – for example, mental retardation, dwarfism, alopecia universalis (no hair) and vitiligo (loss of skin pigment). They also can affect other systems not so readily apparent, such as your hormones, enzymes, metabolism, organ structures and much more. For diseases caused by a single gene defect, the most serious ones will be the earliest targets for genetic engineering cures. We'll get to correction of the other defects later.

For the first several decades of genetic engineering, the tools used to do the actual cutting and patching of DNA were complicated and labor intensive. It took months or years of preparation to set up an experiment for a specific treatment, and then the results were often erroneous. The two most common tools used were called zinc-finger nucleases and TALENs (transcription activator-like effector nucleases). These are still used today, but they have been overtaken by a new tool that has quickly gained worldwide attention: CRISPR.

CRISPR (clustered regularly interspaced short palindromic repeats) is revolutionizing genetic engineering and becoming practically a household name in both the scientific and lay press. If you want to be conversant about genetic engineering, you need to know something about CRISPR.

We owe it all to bacteria. For decades, scientists studying bacterial genes had noticed an unusual sequence of DNA nucleotides in some of them. That sequence is described by the CRISPR acronym. Embedded in that sequence was a stretch of DNA that exactly matched a DNA or corresponding RNA sequence in certain viruses that infect bacteria. (The genomes of some viruses are based solely on RNA and others on DNA.) Further research showed that this match was no accident; it was actually part of a bacterial defense against those viral infections.

Here's how it works. CRISPR consists of two components. The first (usually referred to as the CRISPR component) contains the DNA sequence that corresponds to the target viral DNA or RNA.[142] This generates RNA molecules (called guide RNAs or gRNAs) that move around the bacterial fluids seeking a virus with the matching sequence. Upon encountering such a virus, the gRNA attaches to it. At this point, the second component of CRISPR is activated. This is an enzyme called Cas9 (CRISPR-associated endonuclease-9), which severs the virus genome at that point and thus destroys it. Sometimes CRISPR is referred to as CRISPR/Cas9 for this reason.

It took the brilliance of two genetic researchers – Jennifer Doudna and Emmanuelle Charpentier – and their teams at U.C. Berkeley and The Laboratory for Molecular Infection Medicine in Sweden to recognize CRISPR's potential for genetic engineering. In a paper published in 2012,[143] they showed that if you genetically altered the targeting DNA sequence of CRISPR, you could target *any* DNA sequence *anywhere* for targeted cutting by the Cas9 enzyme.[144]

It is no exaggeration to say that the field of genetic engineering exploded after that publication. You could now set up an experiment to target any gene in any organism in days or weeks, rather than months or years. Start-up kits for CRISPR could be purchased on the internet for $150. You didn't need to be a rocket scientist (or a genetic scientist) to learn how to use it. By the following year, CRISPR/Cas9 was being used "to delete, add, activate, or suppress targeted genes in human cells, mice, rats, zebrafish, bacteria, fruit flies, yeast, nematodes, and crops, demonstrating broad utility for the technique."[145] It could be used on somatic cells, stem cells, germline cells and IVF zygotes. It could be programmed to target multiple genes simultaneously. Companies sprouted up everywhere to attempt to commercialize CRISPR technology. In 2015, CRISPR was named *Science Magazine's* 'Breakthrough of the Year.' Speculations of future Nobel Prizes abounded. The race was on – and it is still going.

We've since learned how to modify the CRISPR/Cas9 system to do more than just cut the targeted gene. We can now, when desired, insert a normal DNA sequence to replace the abnormal cut section. Similarly, a normal copy of an entire gene can be inserted to replace an abnormal gene. This new capability provides the tools to correct any type of single-gene disease-causing mutation.[146] I am simplifying the actual complexity of this procedure here, which will be discussed later relative to the role that AI will play in the future.

But there is more – much, much more. It seems that every month or so, some publication details a new form of CRISPR, or its cutting enzyme Cas9, that shows a new capability or makes an improvement over an old one. That is because there are many different versions of CRISPR/Cas9, in different bacterial species, being discovered and tested.

In addition, known versions are being modified as to how they can be deployed. Some of these variations are in the CRISPR component of the tool (the portion that identifies which DNA sequence to alter) and some occur in the Cas9 portion (the enzyme that does the actual cutting – when this is true, the enzyme may be given a different name, such as Cpf1 instead of Cas9). You may think of CRISPR/Cas9 as a tool such as a knife. Each of these different versions is a different form of the same general tool, customized for special purposes, just like the differences between a scalpel and a bread knife. Table 3 is a partial list of these variations.

NEW OR IMPROVED FUNCTION	VARIATION NAME
Single nucleotide precision editing	CRISPR-nCas9-cytidine deaminase
Multiple gene editing	sgRNA-guided CRISPR
Gene expression inhibition	CRISPRi
Gene expression enhancement	CRISPRa
Improved knock-in enzyme	Cpf1 instead of Cas9
Enzyme targeting RNA viruses	Cas13 – CRISPR/CARVER
Large segment knock-ins	2C-HR-CRISPR
Improved efficiency in bacterial edits	CRISPR/BEST
Jumping gene editing	ShCast V-K CRISPR/Cas12k
'Very fast' precision editing	vfCRISPR
Reduce off-target mutations	anti-CRISPR
Compact editing	CRISPR-CasΦ

TABLE 3 – **VARIATIONS OF CRISPR/CAS9**

There's much more being done with CRISPR, but this gives you an idea of its capabilities and why it's called the Swiss Army Knife of genetic engineering. Its potential is revolutionary, and we are just learning how best to use it. Even at this early stage, it would take a full academic course to teach all of the uses for this tool, in all of its variations. Even then, the course would become quickly out of date because of the continual new discoveries of CRISPR variants and their uses.

Are there any problems with CRISPR? Unfortunately, yes: off-target mutations. These are the same problems we have always encountered in genetic engineering. They also occur in CRISPR-altered organisms. Even more troubling for the future, early experiments on human embryos show this to be a problem. These latter experiments, done on IVF zygotes that are not intended for implantation, have demonstrated that as many as 50% of the embryos edited with CRISPR contained off-target mutations.[147] Much work needs to be done to eliminate this.

Another potential problem is when Cas9 or another CRISPR-related cutting enzyme stimulates the normal DNA repair mechanism. Some laboratory studies have shown that this could stimulate cancer development.[148] Cutting through both strands of DNA (as Cas9 normally does) creates other potential unintended consequences in gene disruptions as well. In addition, we still do not have full control over the precision of inserting new segments into DNA. For example, mosaicism can happen when only some of the target cells are successfully treated while others remain untreated. The impact of this depends on the abnormality. Allergic reactions are also possible. There is much work ahead to make these procedures completely safe and effective.

Then there are the patent fights. U.C. Berkeley, based on Doudna and Charpentier's 2012 published work, filed for a patent on the use of CRISPR for genetic engineering.

Feng Zhang at the Harvard/MIT Broad Institute published a paper in early 2013 on the use of CRISPR in mouse and human cells. Since the Doudna-Charpentier publication involved prokaryotes, the Broad Institute filed for a patent on the use of CRISPR in more complex eukaryote cells, which includes mice and humans. The Broad Institute claimed that since the Berkeley studies were done on bacteria rather than eukaryotes, Berkeley's patent only applied to prokaryotes. The Broad Institute was awarded its patent. Berkeley appealed that decision, claiming it was obvious from their work that CRISPR could be used in any kind of cell, and that the Broad Institute's patent infringed on theirs. Berkeley lost that appeal at both the Patent Trial and Appeal Board level and in the Federal Appeals Court. That should have settled it, but it didn't. In 2019, Berkeley filed additional information to the patent office, which re-opened the issue. The ugly patent fight goes on. The 2020 Nobel Prize in Chemistry, however, went to Doudna and Charpentier.

In the meantime, competing private companies have been formed by both the Doudna group and the Zhang group to try to commercialize various CRISPR applications. There are other participants in the patent melee, and dozens of other CRISPR-related patents have already been awarded to at least 18 different organizations. In the long run, the resulting confusion as to who is allowed to do what (or required to pay whom) may be the biggest problem to overcome in realizing the full benefits of this technology.

On the other side, there are those who consider CRISPR fundamentally dangerous and unethical, as indicated by Bill Joy's quote at the beginning of this chapter. James Clapper, when he was the director of U.S. National Intelligence, called genome editing tools such as CRISPR weapons of mass destruction. In his report in February 2016, he stated the following:

> "Research in genome editing conducted by countries with different regulatory or ethical standards than those of Western countries probably increases the risk of the creation of potentially harmful biological agents or products. Given the broad distribution, low cost, and accelerated pace of development of this dual-use technology, its deliberate or unintentional misuse might lead to far-reaching economic and national security implications."[149]

WHERE ARE WE?

Following the early setbacks and the subsequent genetic engineering 'winter' after the Jesse Gelsinger's death and other fatalities, both the technology and acceptance of genetically modified humans (GMHs) have slowly regained more scientific and public acceptance. This has been especially true with the advent of CRISPR. Animal studies and *in vitro* human cell studies are showing progress in a number of diseases. Many clinical trials are under way around the world and many more are about to begin. Despite all that, the results are still not dramatic. Transferring techniques from mice and other animal models to humans, or from *in vitro* human studies to *in vivo* trials, are always problematic and subject to unexpected complications.

Sickle cell disease affects more than 100,000 patients in the U.S., mostly African Americans. It was the first disease to have its exact molecular basis determined. It is caused by a single mutation that produces hemoglobin in red blood cells. When a patient inherits two abnormal gene copies of this recessive disorder, he or she suffers red blood cell distortions (sickling) that lead to periodically clogged blood vessels. It's extremely painful and causes multiple organ damage and decreased life expectancy. After 70 years of studying this disease, it seemed like an ideal candidate for gene therapy. Yet, after 30 years of genetic engineering experiments in humans, we still do not have the cure. But we're getting tantalizingly close.

The first problem is that mature red blood cells do not contain nuclei (or genes) and therefore cannot be the target of gene therapy. Instead, the precursor stem cells in the bone marrow must be targeted. The first reported genetic treatment of a sickle cell patient was in 2017, in France.[150] Stem cells were removed from her bone marrow and, using a viral vector, were genetically modified to correct the mutation. The remaining bone marrow was treated to destroy the mutation-containing stem cells. Then, the genetically engineered cells were infused back into the bone marrow. The procedure was successful, and the patient has had no further symptoms so far. This study is considered an early proof of concept but requires much larger and longer studies.

A follow-up study of 45 sickle cell patients is now under way in the U.S., using a different protocol. CRISPR will be used to make genetic changes to produce fetal hemoglobin rather than adult hemoglobin. Normally, fetal hemoglobin is produced for a short time following birth, until the production of adult hemoglobin takes over. Fetal hemoglobin reduces the sickling of abnormal hemoglobin when it is present. In July 2019, the first such patient to receive this treatment was Victoria Gray, a 34-year-old mother of four from Mississippi. She is being treated by a team headed by Dr. Haydar Frangoul at the Sarah Cannon Research Institute in Nashville, TN. As reported nine months following treatment, she was living a normal life with no sickle cell attacks and no complications. There are several other clinical trials of genetic engineering cures of sickle cell disease under way, and they are reporting similarly good early results.

We are well on our way to being able to cure sickle cell disease. However, if it requires the type of therapies just described, it will likely not help most of the patients who need it. There are more than 12 million people with sickle cell disease in sub-Saharan Africa, where the medical infrastructure and economics could not support such a complex and costly therapy. For this reason, the Bill and Melinda Gates Foundation announced a joint venture with the National Institutes of Health in 2019 to spend $200 million over the next four years to develop genetic engineering approaches to both sickle cell disease and HIV.

Cystic fibrosis is another common genetic disorder caused by mutations in a single gene. More than 30,000 people have it in the U.S., and since it is a recessive disorder, more than 10 million others are carriers of the disease. The most common form involves alterations of three nucleotides in this gene, although hundreds of other mutations to this same gene are known to cause the disease. Thus, the clinical manifestations are variable, but with good clinical management most patients live into adulthood. Still, this requires considerable efforts and cost, and complications usually shorten lifespans. The worst symptoms involve both the lungs and pancreas, which makes genetic treatment particularly difficult. Most clinical trials have focused on the lung component, for two reasons. The first is that the most serious manifestations are lung infections and other respiratory impairment due to the thick mucus caused by the disorder. The second is that it is much easier to deliver the vector fixes to the lungs than to the pancreas. One can simply inhale them. Although there have been some short-term improvements, the results so far are disappointing. So, unlike sickle cell disease, there seems to be no genetic therapy cure for cystic fibrosis on the near horizon.

Huntington's disease would seem to be ripe for a genetic engineering cure. It is also caused by a mutation in a single gene, but unlike sickle cell anemia and cystic fibrosis, it is autosomal dominant. That means the mutation needs to be inherited from only one parent to cause the full-blown disease. It also means that the abnormal protein is toxic and causes damage to the neurological system. The famous musician Woody Guthrie died of this disorder. The symptoms usually don't appear until ages 30–50, so young people could be screened and treated before any serious neurological damage occurs – if we had a way to knock out the bad gene.

The problem is that the gene is expressed in a certain part of the brain. The capillaries in the brain are different from those in other tissues in that they allow far fewer substances to diffuse into brain tissue – including the viral vectors that might be used to treat brain disorders. This so-called blood-brain barrier is a natural protective mechanism for the brain. After many years of research and animal testing, a clinical trial using a common viral vector has been started. This will insert a gene to permanently suppress the expression of the mutated gene. Unfortunately, this requires that the corrective DNA be injected directly into the brain tissue during neurosurgery. So, even if this treatment is successful, the major neurosurgery requirement remains problematic. Therefore, the best solution would be to prevent the disease by strongly urging pre-implantation genetic diagnosis (PGD) and IVF in any cases where there is a family history of Huntington's disease. That will not prevent all cases, however, so treatment trials are still necessary.

Genetic engineering is having early success in a number of other diseases, most of which are not well known or highly prevalent. Without going into the details, a partial list of diseases shown in Table 4 below have had some long-term success in clinical trials. They all involve correction of a defect in a single gene. Although the exact mutation may vary from patient to patient, once the genetic abnormality is determined, the correction can be directed toward that specific defect. These treatments are not yet considered cures because they need longer periods of follow-up and larger numbers of patients, but they do indicate that many have a chance to succeed within the next decade.

DISEASE	PREVALENCE	GENETICS	CLINICAL	COMMENTS
Sickle Cell Disease	100,000 in U.S. 12M in Africa	Autosomal recessive	Pain, organ damage, shortened lifespan	See text on previous pages
Cystic Fibrosis	30,000 in U.S. (>10m carriers)	Autosomal recessive	Lung, pancreas problems, shortened lifespan	See text on previous pages
Beta Thalassemia	1,000 in U.S., higher in malaria endemic countries	Autosomal recessive	Severe anemia	CRISPR Therapeutics of blood stem cells to increase fetal hemoglobin
Wiskott-Aldrich Syndrome	1/250,000 male births	Sex-linked recessive	Bleeding, infections, skin problems	Lentivirus vector Rx of blood stem cells
Adrenoleukodystrophy	1 per 21,000 males	Sex-linked recessive	Neurological and adrenal problems	Lentivirus vector Rx of blood stem cells
Hemophilia A and B	A - 1/5,000 males B - 1/30,000 males	Sex-linked recessive	Bleeding	Gene Rx viral vector targeted IV to liver cells. Multiple clinical trials – good results up to 3 years.
Leber congenital amaurosis – type 2	1/40,000 newborns	Autosomal recessive	Blindness	Multiple trials. Direct injection of viral vectors into retina. Direct CRISPR injection also used.
Junctional epidermolysis bullosa	0.49 cases per 1m live births	Autosomal recessive	Severe skin disorder, shortened lifespan	Can now regenerate normal skin with gene Rx
Hunter syndrome	1/125,000 males	Sex-linked recessive	Nervous system and other organ failure	Infusion of liver with viral vector restores normal enzymes
Spinal muscular atrophy	1/9,000 worldwide	Autosomal recessive	Paralysis and early death	Infusion of viral vector crosses blood-brain barrier and corrects defect. FDA approved Zolgensma in 2019.
Severe combined immunodeficiency	1/75,000 male births	Sex-linked recessive	'Bubble boy' syndrome – severe infections	Infusion of self-inactivating lentiviral vector
Huntington's disease	1/15,000	Autosomal dominant	Neurological problems and shortened lifespan	Trial under way using viral vector direct injection into brain tissue for suppression of gene expression

TABLE 4 – **EXAMPLES OF EARLY GENETIC ENGINEERING ATTEMPTS**

Again, this table is just a partial list to indicate the variety of diseases that have already shown encouraging results. To be complete, it would have to include about 1,000 trials presently ongoing. There are also many more diseases in pre-clinical research with clinical trials likely to begin shortly.

GERMLINE GENETIC ENGINEERING

As discussed earlier, your body has two types of cells: somatic and germline. The overwhelming majority of your cells are somatic, which means that any genetic alterations of them are not passed on to future generations. All of the clinical trials for the diseases mentioned above, including cancer, involve somatic cells. Germline cells are the sperm in males, the eggs in females and the stem cell precursors to them. The fertilized eggs, or zygotes, created during IVF are also germline cells. Just as any mutations in germline cells are passed on, so too would be any genetic engineering alterations in those cells.

In the U.S., and most countries, there is a ban on germline genetic engineering in any patient or in any IVF zygotes that are intended to be implanted. However, that does not apply to experimenting with IVF zygotes that will be discarded prior to implantation. This research provides valuable information on the best techniques and discovering possible complications, including off-target mutations. Such research began in China, but it has now spread to the U.S. and other countries. We are learning the best ways to alter these germline cells with the least amount of risk, in preparation for the eventual day when germline genetic engineering will be allowed. Yes, one day we will have genetically modified humans (GMHs) resulting from germline genetic engineering - certainly in China and perhaps elsewhere. My prediction is that it will eventually be allowed in the U.S., but only after other countries take the lead. And, as you will see in the next chapter, it will be a requirement to achieve the *Fourth Great Transformation*.

In fact, a form of germline genetic engineering has already been performed on a small number of patients in the U.S. Remember from Chapter 2 that the cells of all eukaryotes (including humans) contain mitochondria that originally were engulfed bacteria that became part of the eukaryote cell. Those bacteria themselves had genes. Most of those genes eventually migrated into the nuclear DNA, but a small number - 37 out of our 20,000-plus genes - remain in our mitochondria. Those genes have important functions, especially relating to energy metabolism. Mutations in them can lead to the inability to sustain a pregnancy as well as cause other serious genetic diseases.

In the 1990s, Dr. Jacques Cohen at the St. Barnabas Institute in New Jersey pioneered an IVF procedure to prevent a mother with a mitochondrial genetic disorder from passing it on to her offspring. Mitochondrial genes are only passed from the mother, not the father. The technique involves removing the nucleus from one of the mother's eggs and implanting it in the egg of another donor woman whose nucleus had been removed. This creates a new egg that contains all of the nuclear genes of the mother, combined with only the normal mitochondrial genes of the donor. That egg is then fertilized with the sperm from the intended father. Figure 8 is a diagram of this procedure.

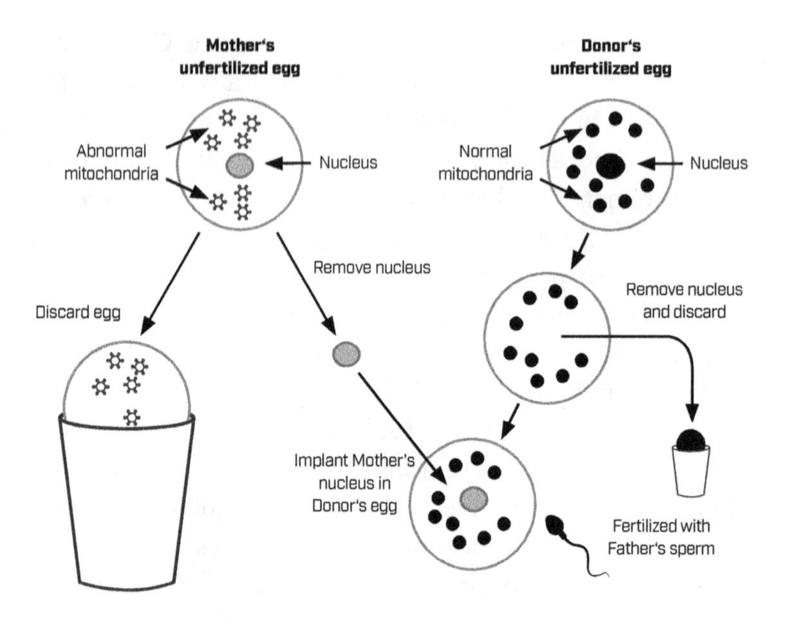

FIGURE 8 – **MITOCHONDRIAL GENE THERAPY**

The resulting zygote is then implanted in the mother. The child will have virtually all of its inheritance – the 20,000-plus nuclear genes from its true parents and only the 37 mitochondria genes from the donor. These mitochondria genes are free of genetic disease and will be passed on to future generations. In essence, this child has three genetic parents and this procedure is certainly a form of germline genetic engineering.

By the early 2000s, as many as 50 patients worldwide had successfully undergone this procedure, performed by Dr. Cohen and others. However, one child developed an autistic-like cognitive condition. It was unknown if that condition was the result of the procedure. The FDA, in an overabundance of caution amidst strong public disapproval of germline therapy, instituted an onerous approval process for this treatment, which essentially ended it in the U.S.

Later, the FDA banned all germline therapies, which is still the status today.

It was not the end of this procedure, however. There has been one reported case where an IVF specialist in New York City, Dr. John Zhang, performed the procedure for American parents by doing the actual zygote creation and implantation in Mexico, where it is allowed. It's unknown how many others have done the same thing. In 2015, mitochondrial gene therapy became legal in Great Britain. This procedure is apparently also being performed in Russia, Ukraine, Spain, Albania and Israel without any official recognition of its legality. Since more than 12,000 women in the U.S. alone are known to have some form of mitochondrial disease, there are lots of options for medical tourism. Keep in mind, though, that this treatment only prevents a mitochondrial-based genetic abnormality from being passed on to offspring. It does not deal with people already living with those diseases. Curing them would require true genetic engineering.

It could be argued that mitochondrial gene therapy as described above is not truly genetic engineering. There is no recombinant DNA involved. It is simply taking mitochondria with normal genes from a donor woman and substituting them for another woman's mitochondria with diseased genes. So, it is really just a mitochondrial transplant. What is not allowed anywhere in the world today is purposefully altering any germline cell genes in a person or in an IVF zygote intended for implantation. We simply don't know enough today to do that safely. Furthermore, the ethical considerations and consequences are far from agreed upon.[151]

BOMBSHELL!

In November 2018, the news broke that twin girls in China had been born the previous month after genetic engineering had been performed on their fertilized eggs prior to implantation in an IVF clinic. This had been done by a biophysicist named He Jiankui at the Southern University of Science and Technology in Shenzhen. Jiankui had used CRISPR to knock out the CCR5 gene that normally produces a surface protein on human immune cells. This protein is what the HIV virus uses to enter these cells and establish the disease. Some people have a natural mutation that prevents this gene from producing the surface protein and, consequently, they are resistant to HIV. It was assumed that knocking out this gene to stop production of the surface protein would also provide HIV immunity.

All hell broke loose immediately following the announcement. Almost the entire scientific world condemned Jiankui for doing this work, calling it premature, reckless, unnecessarily risky and totally irresponsible. Commentary in the press referred to Jiankui as China's Frankenstein. Although immunity from HIV is a worthwhile goal, it is hardly a life-threatening genetic disorder justifying this type of risk at this stage of our knowledge of human germline genetic engineering. Jiankui was dismissed from his university position and has been ostracized by former colleagues, who have distanced themselves from the work. Some of them had been aware of his plans and, at least now, claim they tried to discourage him from doing the procedure. In any case, they didn't inform the authorities. Adding to the controversy, a second pregnancy with the same modification by Jiankui was under way at the time. There has been no further news about this child, who would now be over one year old. On December 30, 2019, Chinese state media reported that a court in Shenzhen had sentenced Jiankui to three years in prison for conducting "an illegal medical practice."

As far as we know, the twin girls are normal. Their father was being treated for HIV throughout this period, although fear of the offspring being infected was not Jiankui's motivation. There are safe ways to protect IVF zygotes, the subsequent fetus and the newborn from HIV infection transmitted by infected male sperm. Rather, Jiankui said the family was suffering from the stigma of HIV, so his and their motive was to spare the children that same fate. Jiankui claims the parents were informed of the risks and approved the procedure. He said that the hospital ethics committee fully reviewed and approved it as well. The hospital denied this and claimed the approval forms had been forged. In addition, later analysis of the forms and procedure information given to the parents concluded that "the participants in this study were clearly misinformed about the study's purpose, as well as being subjected to considerable pressure (via free IVF) to take part."[152]

Only time will tell how these children will be affected. It was reported that there is evidence in Jiankui's unpublished data that the gene editing was not uniform throughout the embryos prior to implantation.[153] That would mean the children are mosaics, and that some of their bodies' cells have the genetic alteration and some do not. It would then be unclear whether they really have HIV immunity, or if there may be other partial cell alterations or off-target mutations. So far, we have no further information regarding the three children involved.

So, what's the status of germline genetic engineering? Virtually the entire world, including China, has now prohibited it. On the other hand, most countries, including the U.S., are still allowing further research on human embryos not intended for implantation.

Another genetic researcher has already announced his intention to implant genetically altered IVF zygotes. Denis Rebrikov, who works at Russia's largest fertility clinic, in Moscow, initially said he would perform the same edit as Jiankui on the CCR5 gene using CRISPR. His target patients would be mothers with active HIV who are not responding well to treatment. Such women have a higher chance of passing on the virus during pregnancy. However, later reports said that he would target, instead, a gene causing deafness. The regulations against germline genetic therapy in Russia are not well defined. Rebrikov claims he will apply for official government approval before performing any procedure, which now seems unlikely to be obtained. As of this writing, there are no reports from Russia of any germline therapy actually being done.

What jumps out at me about the Rebrikov publicity is that he is considering curing deafness as a demonstration of germline therapy. Deafness? Really? It is hardly in the category of Huntington's disease, hemophilia, OTCD or other life-shortening genetic illnesses. We already have other treatments for deafness, such as the cochlear implants discussed in the previous chapter. And some would argue that deafness is not even a disease that necessarily needs a cure. Deaf people have their own culture and society and lead relatively normal lives. To some, the idea of 'curing the deaf' is a form of discrimination.

It is inevitable that germline genetic therapy during IVF will be become legally available somewhere in the world. In some perverse way, the Jiankui incident may have hastened that day by bringing the discussion into the open. There has been little outcry over the mitochondrial germline genetic engineering now approved in Great Britain. Although that is quite different, the nuances are lost on much of the public.

We are currently unable to prevent potential issues from off-target mutations. Even the on-target effects of altering a single gene are not fully known. For example, studies in mice have shown that the CCR5 gene targeted for HIV resistance also seems to impact intelligence, strokes and other neurological factors. Other studies indicate that the normal gene form may improve resistance to other infections, even though it lowers resistance to HIV. The complexity of our genome is so great that it will take AI to achieve a higher level of competence before safely altering it. But with AI advancing as rapidly as genetic engineering, that does not represent a barrier.

WHY AI?

The answer to this question is simple: 6 billion. The human genome consists of 3 billion pairs of nucleotides. We know that a change in just one of them in the wrong place can spell the difference between life and death. We talked about the use of AI for the game of Go in Chapter 4. Go has 10^{320} possible moves, which is greater than the number of atoms in the universe. Well, that number is small compared to the number of possible versions of the human genome, the proteins it produces and all the various combinations. One wonders how we can even now be manipulating the genome.

Currently, we are only focusing our genetic engineering on a tiny portion of those 6 billion nucleotides. We are now only trying to correct single-gene-causing diseases. There are more than 10,000 of those, but we are concentrating on perhaps a few dozen for the foreseeable future. That brings the numbers of nucleotides we are dealing with down to a manageable total, considering our available technology and computing power. Nonetheless, there are still problems and risks associated with what we are doing.

The problem is that we don't fully understand everything we're doing. First and foremost, we know that every technique, including all of the CRISPR versions, have the potential for off-target mutations anywhere in the genome. Just being able to look for them takes massive computing power. Second, we may know the relationship of a given gene with a given disease, but we do not always know what else the protein produced by that gene may be doing elsewhere. We discussed earlier the example of the CCR5 gene for HIV vulnerability also being involved in other infection susceptibility and neurological functions. The sickle cell gene helps protect against malaria. The cystic fibrosis gene mutation may be protective against tuberculosis. We have many such examples of gene interplay without even looking for all of the possibilities. Also, the same gene can produce more than one protein, or RNA molecules, that could affect gene expression. We know neither what all of them are nor what they do.

Truly understanding all of this takes more than just looking at millions of genomes. It requires looking at myriad phenotypes and how they relate to the multitude of genotypes. It will certainly require the assistance of AI to discover and sort out the hidden patterns in these relationships. The more we understand these patterns, the more safely we can attempt to alter our genome. Our databases are growing exponentially, and it is estimated that we have generated more data in the past two years than in all of previous human history. The number of human genomes sequenced to date is in the millions, and will likely reach more than 2 billion within ten years. Similarly, our computing power is growing exponentially and, with the likely onset of quantum computing, will take even further leaps. However, neither the data nor the AI power is sufficient today. We can accept errors in game playing, but our tolerance for error in genetic engineering is far less.

The diseases we are dealing with now are so deadly that we're willing to take some risks. We perform careful clinical trials to try to discover all the unknown consequences.

However, these studies can only enroll so many patients and they can't represent all the variations found in nature. Thus, we know our current protocols are far from complete.

Now, add to this all the next steps that will surely come. We will develop diagnostic and treatment approaches to the most common diseases that will involve multiple genes, perhaps even thousands. This will require examining millions of locations on the genome and correlating each one to a particular disease. That correlation will require studying millions of genomes and similar numbers of phenotypes. Our studies need to expand both the numbers of people and the number of areas on the genome to be examined. It will certainly require AI to make sense of all that data. Only AI can handle the number of simultaneous variables that will be needed in evaluating multi-gene genetic engineering. That will also be true when progressing to single-gene modifications for either disease or enhancement that we don't even consider doing today because the potential benefits do not warrant the risk. Finally, moving to germline genetic engineering will require an entirely new level of risk reduction to be accepted.

In summary, AI will be required to bring genetic engineering to a point where it is safe enough to be applied to complex genetic disorders, human enhancement and, specifically, to germline genetic engineering. It is not entirely predictable exactly what those AI contributions will be any more than it was predictable what were the best moves in chess, Go, poker or other games, or in all the other areas to which AI has already made contributions. AI is what guides us through complexity, and our genome – along with the human brain – is one of the most complex structures in existence.

GERMLINE GENETIC ENGINEERING IS INEVITABLE

Germline genetic engineering is coming. Why? The simplest answer is that genetically engineering IVF zygotes is easier and safer than genetically engineering germline cells in a person where the treatment must be delivered to the appropriate cells. That problem does not exist with IVF zygotes since no complicated vector delivery is required. Those altered genes will exist in every cell, and the normal epigenetic controls on gene expression will ensure that the appropriate cells express the appropriate genes, even the altered ones.

For example, we discussed the difficulty in genetically treating cystic fibrosis because the abnormal gene is expressed in multiple tissues. That dramatically complicates performing somatic genetic engineering on living humans. Germline genetic engineering of IVF zygotes, however, solves that problem and allows normal children to be born from two parents with the disease. If there is an error in the procedure, though, it would affect all future generations.

BRAVE NEW WORLDS

In this chapter, we have been discussing genetic engineering of genomes that have been the basis of all life for billions of years. These fundamentals of DNA and genes have been stable over this entire period. *Homo sapiens* are now changing that by re-engineering these very fundamentals of life. What follows are some examples of how we are doing that.

GENE DRIVE

Meiosis. That's how most of us eukaryotes reproduce. We have sex. A male and a female get together and each contributes half of the DNA needed to produce the child. There are two copies of every gene in the fertilized egg and in every cell that develops thereafter – one from the mother and one from the father.

Until now, that is. After more than a billion years of this fine sexual tradition, we have found a way to change that, supposedly for our own good. The genetic engineering technique is called *gene drive*. Normally, when a sperm or an egg is created, there is a 50/50 chance that the gene copy in the sperm or egg will come from either one of the two copies each of us has of that gene. With gene drive, one of those copies is genetically engineered so that up to 100% of the time it will be the copy that is passed along. The same thing would happen in each subsequent generation, so that over just a few generations, the genetically engineered gene will prevail throughout the population.

This is difficult to understand. Think of it in terms of coin flipping. Rather than nature performing its usual coin flip to determine which of two copies of each gene gets into the sperm or egg (that is, from the original mother or father), it instead acts like a two-headed coin and the result always comes up heads. The gene that contains the gene drive alteration always enters the sperm or egg. The figure below illustrates this for four generations.

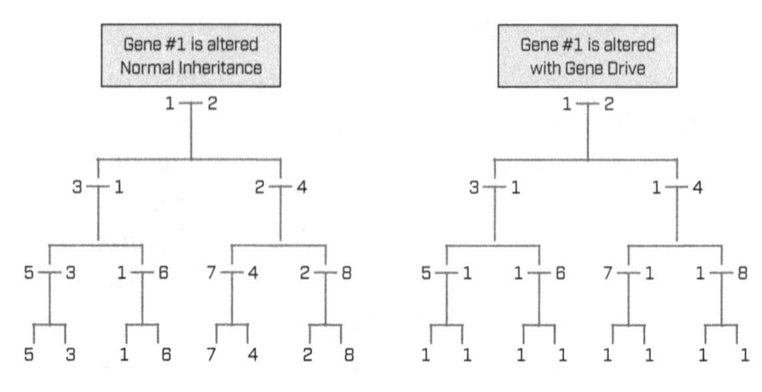

FIGURE 9 – **GENE DRIVE**

Note that if the gene drive alteration in gene #1 causes infertility, then over some number of generations that species becomes extinct.

Why would anyone want to do this? Malaria, Zika, Dengue Fever and Yellow Fever are each spread by a specific species of mosquito. Prior to gene drive, genetic engineering had already been used to reduce the population of disease-carrying mosquitos. This was done by genetically altering male mosquitos in the laboratory to either be infertile or to cause lethal genes in their offspring. Releasing huge numbers of these genetically altered males into the wild dramatically reduces the number of mosquitos in that area. However, the effect is only temporary and requires the continual release of the altered males.

Gene drive makes that permanent by totally eliminating a given species. It causes that particular genetically altered gene to be passed on to most, if not all, subsequent fertilized eggs, rather than just one-half of them in the normal case. In laboratories where this has been done, the mosquito species becomes totally wiped out over a small number of generations. If we do that to the one species of mosquito that spreads malaria, we could eradicate a disease that causes millions of deaths. Cheap. Quick. No need to develop a vaccine and spend millions trying to vaccinate everyone.

Gene drive is controversial, to say the least. We have never deliberately tried to wipe out an entire species of anything by genetic engineering. What effect would there be on the ecology if an entire species of mosquito was eradicated? What other animals depend on that species as their main food source? On the other hand, since there are thousands of species of mosquitos, maybe it wouldn't have any negative effect. On the *other* other hand, what if a terrorist group created a gene drive to spread a deadly plague by some insect? Or what if we make a mistake in the gene alteration and it inadvertently creates some other problem? The risks seem incalculable.

Scientists, regulators and the interested public have been agonizing over gene drive for years. Should it be banned? Should it be regulated? Who should approve any deployment? Who owns a species? Mosquitos don't stick to national boundaries. Could the U.N. control this? And note that gene drive can be done in any species – even humans.

Because of these concerns, the first release of gene drive mosquitos in the wild is probably years in the future – if it ever actually happens. At least that's the plan if the project Target Malaria controls that release. The not-for-profit research consortium's goal is to eradicate malaria, and it has a deliberate, phased approach to studying the use of gene drive in mosquitos and its potential deployment in Africa. To quote from the project's website:

> "Target Malaria started as a university-based research programme and has grown to include scientists, stakeholder engagement teams, regulatory affairs experts, project management teams, risk assessment specialists and communications professionals from Africa, North America and Europe. Target Malaria receives core funding from the Bill & Melinda Gates Foundation and from the Open Philanthropy Project Fund, an advisory fund of Silicon Valley Community Foundation."

The project plan involves three phases, the first two of which do not involve gene drive. The first phase has already begun, with the first release of genetically modified mosquitos into the wilds of Burkina Faso. These mosquitos are all sterile males that will die out naturally without producing any offspring. They are also coated with fluorescent dust that will allow researchers to follow and carefully track their dispersion, behavior and outcomes. To illustrate their deliberate approach with this first cautious release, this phase has been in the planning stages for seven years. The second phase will be to release a new batch of genetically modified mosquitos that are fertile but only produce male offspring, which are modified to die out in the wild. Researchers will study these two phases, which will take more years and will include total transparency with all stakeholders. Only then will the project proceed to the third phase release, which will include the gene drive. That's assuming there are no problems or issues in the first two phases.

Gene drive is still years away from deployment. Of course, that assumes there will be no rogue implementations anywhere else in the world. In the meantime, there have been new developments in genetic engineering techniques that could be used to reverse a gene drive even after organisms have been released into the wild.[154] That could allow earlier real-life experimentation.

CLONING

Remember Dolly, the sheep in Scotland? In 1996, she was the first cloned mammal anywhere in the world. Dolly was not a mixture of a mother and a father but was an exact genetic duplicate of another sheep.

Scientists took the nucleus out of a mature adult cell of a donor sheep – the one they wanted to duplicate – and put that nucleus into a sheep egg that previously had its nucleus removed. Normally, mammalian eggs in the ovaries have only half of the normal complement of genes since they will later be fertilized by a sperm to complete the full set. But in this case, the egg was given an adult nucleus from the donor sheep, so it already had a complete set of genes, all from the donor sheep. It did not need a sperm. Then the new egg was stimulated electrically to induce it to begin developing in the same way a fertilized egg would develop. When it reached a certain stage, it was implanted into the uterus of another sheep, which carried it to full term. The result was Dolly, a clone of the donor sheep.

Dolly had normal offspring of her own, but only lived to about 6.5 years old. She had arthritis and died of a lung cancer that is common in sheep. That is only about half of the usual sheep lifespan, but her donor DNA was already six years old, which could have accounted for that. We have since cloned thousands of mammals, including pigs, deer, horses, bulls, cats, dogs and even more of Dolly. There is no evidence that cloning causes any more or less than the usual diseases or aging of any species.

Boyalife is a Chinese company that clones beef cattle for the Chinese market, among other genetic engineering endeavors. Its goal is to produce a million cloned cattle annually. A South Korean company, Sooam Biotech, will clone your dog, alive or recently dead, for a fee. A Chinese company called Sinogene will do the same thing. Cats, too. Barbra Streisand cloned her beloved pet Samantha, a Coton de Tulear dog. Actually, twice, reportedly costing her about $200,000.

In 2018, the first primate species was cloned in China – a macaque monkey. Theoretically, that could bring us closer to cloning a human, but there is no intention by anyone in China or elsewhere to do that, and it is banned in the U.S. Instead, the intent is to clone monkeys with specific genetic diseases or other characteristics that could be used for research into human disorders. Monkeys are more closely related to humans than mice and other laboratory animals.

In 2009, cloning was used for the first time to attempt to bring back an extinct species – the Pyrenean Ibex, a type of mountain goat. Skin from one of the last remaining members of this species had been preserved in liquid nitrogen since 2000. Spanish scientists thawed some of these cells and extracted their nuclei, to implant in domestic goat eggs and subsequently into the uterus of a goat. One of these carried to a live birth, but the newborn Pyrenean Ibex died within minutes. Still, it raised the hopes that eventually this technique will be perfected to bring back other species whose DNA might be preserved in cold environments, such as the woolly mammoth. Plans call for elephants to be used as carriers for such future experiments. Others are planning to try to bring back the aurochs, the extinct ancestor to modern day cattle. All this has even raised speculation that one day we might be able to bring back a Neanderthal, since we now can produce a copy of its genome.[155]

BRAIN STUDIES

In Chapter 5, we saw that brain studies were key to achieving the goal of the Kurzweil singularity. What wasn't discussed then were the myriad ways that genetic engineering could be used in the study of the human brain.

As mentioned above, and described in Chapter 2, the complete genome of the Neanderthal is now known. With the help of CRISPR, scientists are now growing a Neanderthal brain.[156] It's not a complete brain, being only a few millimeters in size. Brain cells with early brain organization were used from several genes from the Neanderthal genome. Scientists used CRISPR to modify *Homo sapiens* stem cells *in vitro* with these brain-related Neanderthal genes and are letting them grow. The purpose is to study how they form neuron networks and synapses, and compare that to how our own brains develop. The goal is to understand what differences, if any, there are between the two species regarding brain development.

That's not the only attempt to push the envelope. As noted earlier, scientists in China have genetically engineered macaque monkey embryos to add a human gene important in human brain development.[157] They then tested those grown monkeys and found that they actually did perform better on some mental tasks, such as short-term memory. Encouraged by this success, the Chinese researchers are planning to add other brain-related human genes to monkeys in further studies. In another study done jointly by researchers in Germany and Japan, marmoset monkeys were genetically engineered to express a gene found only in humans that enlarges the neocortex.[158] These monkey fetuses, that were terminated prior to birth, had larger than normal neocortices indicating a likely evolutionary path that resulted in the large growth of human brains.

The purported purpose of such studies is to better understand how the *Homo sapiens* brain can do so much more than the brains of other related species. Although I too would like to understand how the *Third Great Transformation* came about, others and I question both the science and the ethics of this approach. Does such genetic manipulation really tell us what happened in evolution? There are millions of genetic differences between monkeys and humans that occurred over 25 million years in countless environments. It seems like random chance to select only a few of those genetic differences for study in a laboratory setting that is totally unlike anything that occurred in nature.

More importantly, what gives us the right to inflict these kinds of studies on monkeys or any other animals?

Just to satisfy our curiosity? What if we really are making these monkeys more human-like? If so, that might seem less like research and more like torture.

DNA PRINTER

We can now create in a laboratory any DNA sequence we desire, utilizing a special type of printer. There are commercial companies that perform this to order, for scientists, drug companies and anyone studying genetics or genetic engineering. These companies create small segments of DNA to be used as recombinant DNA, and could create an entire gene if the sequence is known. That could save them the trouble of obtaining an actual organism to extract a normal copy of a gene. For humans, a normal copy of a gene could be created and inserted into a patient with an abnormal copy. Someday, this technology could create one of those human artificial chromosomes discussed earlier, with precisely the desired new genes.

In 2007, the J. Craig Venter Institute synthesized the entire genome of a bacterium.[159] The process is totally controlled by computers, but the actual DNA sequences created are biological. These sequences can be inserted into bacteria to produce large quantities of any given protein. They can be used to study possible new drugs and be used in many ways in the genetic engineering of plants and animals, including humans.

These DNA synthesizers, as they are called, are not exactly like computer printers, but are analogous to them. The desired DNA sequence is created on a computer and sent under computer control to a special device. This device selects nucleotides in a chemical solution and strings them together in a liquid medium – similar to how a printer sprays ink onto paper. The DNA molecules so created can then be replicated chemically, many times over, for their intended use.

The most ambitious project to date is an attempt to create the entire 16-chromosome genome of the brewer's yeast fungus called *S. cerevisiae.* This project is called Sc2.0 and is headed by Dr. Jef Boeke of the Johns Hopkins School of Medicine. It will be the first time an entire synthetic genome of a eukaryote organism is created. Many international organizations will collaborate in creating segments of the full genome. The project's goal, as stated on the team's website, is:

> "...to answer a wide variety of profound questions about fundamental properties of chromosomes, genome organization, gene content, function of RNA splicing, the extent to which small RNAs play a role in yeast biology, the distinction between prokaryotes and eukaryotes, and questions relating to genome structure and evolution. The availability of a fully synthetic genome will allow direct testing of evolutionary questions not otherwise approachable."

Initially, these DNA printers - or, *artificial gene synthesizers* - were used to reproduce known DNA sequences found in nature. But they are now being used to try to improve upon nature. For example, there is redundancy built into the codon, or three-letter nucleotide code, for translating DNA into specific amino acids. That means that for most amino acids there is more than one three-letter code that leads to that amino acid being incorporated into a protein. It's now known that the rate of production of a given protein varies depending on which code is used. So, one use of this process is to change the DNA sequence to optimize the production rate. There are other more complicated (and more difficult to explain) optimizations that can also be performed.

Could these DNA printers be used to produce something bad? Of course they could, if the DNA sequence of the bad thing was known. In 2017, a virologist in Canada synthesized the virus that causes horsepox – a relative of smallpox that does not cause disease in humans. Both horsepox and smallpox have very large and complicated DNA structures, and there had been some question whether either of them could be synthesized in a laboratory. Now that it's possible, and that the DNA sequence of the smallpox virus is known, we know that something so terrible as smallpox could be created in a laboratory with the right expertise. It would be illegal to do that in the U.S., but legality doesn't apply to terrorists. It was not illegal to synthesize the horsepox, although some argue it was not a good idea. It would also not be illegal to take a mostly non-communicable virus and make it more communicable. Although there are national and international organizations that have rules and recommendations regarding genetic engineering, there is no international enforcement.

When I discussed Darwinian evolution in Chapter 1, I said that our problem in genetic engineering is not whether we have a good typewriter, but rather knowing what to type. We now have a good typewriter. We're also getting better about knowing what to type, but as I said, there will be no *Homo deus*. I was referring to Yuval Noah Harari's book, *Homo Deus*, in which he suggests we will have 'godlike' capabilities in the future. Although some will say that much of what I'm describing here is already godlike, it is still a long way from what nature has been able to do over billions of years. No one today could use an artificial gene synthesizer to create a gene to perform a function that has never occurred in nature. For example, we could not create a new type of kidney, or a skin sensor for brain waves, or a single gene that would give someone genius-level intelligence. We simply wouldn't know what DNA sequence would lead to what kinds of proteins to do those things. We can improve on nature only by rearranging things already residing somewhere in a genome.

But we have been trying. We are beginning to change the fundamentals of biology in ways that were unimaginable only a few years ago.

ALIENS

In 2012, Harvard geneticist George Church and science writer Ed Regis published the book *Regenesis: How Synthetic Biology Will Reinvent Nature and Ourselves.* In this eye-opening description of the emerging powers of genetic engineering, they describe the new discipline of synthetic biology as "the science of selectively altering the genes of organisms to make them do things that they wouldn't do in the original, natural, untouched state." We now have the ability to describe any DNA sequence in a computer, translate it into actual biological DNA components, and insert them into organisms. Thus, we can essentially manufacture synthetic organisms with any feature that already exists somewhere in nature.

As examples of future possibilities, they described putting the elephant genome in a computer and selectively altering it in the right places to look like a mammoth, using the known mammoth genome as a template. Then, create the biological DNA, insert it into an elephant egg using cloning techniques, and implant it an elephant, which would give birth to a mammoth. They similarly speculated on a sequence to convert *Homo sapiens* DNA into Neanderthal DNA.

Although ethical and other considerations may never allow such experiments to be done, in 2010, a team at the J. Craig Venter Institute did the same thing with bacteria.[160] They synthesized the entire genome of one bacteria species called *Mycoplasma mycoides*. They then removed the entire DNA from another species of bacteria called *Mycoplasma capricolum* and inserted the synthesized genome into it. Voila, the former *Mycoplasma capricolum* bacteria, lived and reproduced generations of bacteria that were genetically the same as *Mycoplasma mycoides*. This is similar to how a virus takes over parts of its host's biology to recreate more viruses. However, in this instance, the synthetic genome took over the entire reproductive machinery of its host and caused it to make a different species of bacteria. The inserted genome was a synthetic copy of the genome of another species, but everything else in the host was the original bacterium.

That this worked was truly remarkable, and it set the stage for an entire new world of synthetic biology. How did the Venter team know the synthetic bacteria produced were really from their transplanted genome and not just some contaminant? Their proof was equally amazing. They had inserted what they called watermarks into the synthetic genome. DNA is simply a code of nucleotides containing information for creating proteins and RNA. So, this same code could be used to encode textual information simply by creating a translation table for combinations of DNA that translate into letters of the alphabet. Inserting such messages into non-functional areas of a genome would do no harm, but could be used to store messages of all types, as long as we're able to synthesize any DNA sequence.[161] They could also be used, as the Venter team did, to verify their genomes.

The researchers inserted the following messages into their synthetic bacteria:

> "To live, to err, to fall, to triumph, to recreate life out of life."
> "See things not as they are, but as they might be."
> "What I cannot build, I cannot understand."

All famous quotes. They also inserted the names of all the researchers involved in the project. Whimsey? Perhaps. But it was also useful to verify a synthetic organism, provide patent protection for patented synthetic organisms, and for other uses.

Another example of where synthetic biology is heading is a remarkable paper published in January 2020 about the creation of 'living robots.' A team of researchers used a complex computer simulation on a supercomputer to design possible three-dimensional versions of an organism created from frog stem cells for both frog skin and heart cells. They took the most promising designs for a certain function, actually synthesized the configurations of living frog cells in a petri dish, and then observed the behavior of these tiny living creations. These were living organisms less than one millimeter in size, but entirely new life forms. Although they have no present use, the researchers envision future living robots that will be created for drug delivery in humans, internal surgery, removing plaque from arteries, attacking cancer, cleaning toxic waste from the environment and multiple other functions.[162]

These synthetic organisms are truly different from anything that exists naturally. But is synthetic the same as alien? Most people think of aliens as those weird-looking beings in *Star Wars* or other sci-fi movies. But every one of those weird-looking creatures could consist of some variation of the same DNA, nucleotides and amino acids as creatures we already have on Earth. Ever since life began on this planet, all living things have been comprised of these same components, no matter how varied that life has been and continues to be. It has been the most amazingly stable aspect of evolution.

But now, for the first time, we are creating true aliens that do not consist of the same components as existing life forms. As reported in the *Proceedings of the National*

Academy of Sciences,[163] researchers have successfully added two new nucleotides, labeled X and Y, to the genome of *E. coli* bacteria. These new Xs and Ys became integrated with the usual A, T, C and G to form a six-nucleotide DNA that is stable and gets passed on to future generations. The publication reported that these new Xs and Ys didn't do anything, but they also didn't do any harm. That is, this strain of *E. coli* still produced the same proteins it always did, with the same 20 amino acids it always used. This was a first big step to greatly expanding the potential to add new components to DNA, which could add new amino acids and ultimately new proteins into life. This was the first step, then, toward creating completely new forms of unnatural organisms. They would be true aliens, in my view.

The researchers using the X and Y nucleotides did not stop there. By the end of 2017, they demonstrated that they could be used to predictably incorporate new, unnatural amino acids into the proteins of this *E. coli* strain. They engineered two new codons in order to do this. That is, they defined three-letter codons that went beyond the use of the normal A, T, C and G to include either an X or a Y. This would be like adding two new letters to our alphabet to create words that didn't exist before, and then using those words in sentences. This analogy is strained because we would have difficulty understanding a word or sentence that contains new letters beyond our normal 26 – especially if the newly spelled words were intended to have new meanings. That's exactly what's now being done with the genetic code, but organisms are actually understanding the new DNA alphabet.

Why limit this new DNA alphabet to six letters? Another group has shown they could create DNA with eight letters (nucleotides) that they call *hachimoji DNA*.[164] It is speculated that there could be many more viable variations of the basic template of life. Who knows what alien organisms we may find one day, somewhere in the universe, with totally different DNA? And how different does it need to be before we stop calling it DNA? Do we need to create a new definition of life?

Why would anyone want to create synthetic organisms containing unnatural DNA, unnatural amino acids and unnatural proteins? The reason is to demonstrate that these new amino acids and the proteins that contain them have new properties that could improve their functions. The potential, as yet unrealized, is to develop new capabilities in organisms, new drugs for humans and new substances for myriad uses. The potential is enormous, and at this time the outcomes are totally unknown. Ours is the bravest of brave new worlds.

THE FOURTH GREAT TRANSFORMATION?

Will *Homo nouveau* be a product of this brave new world? It could, but in my view that will be unlikely. Virtually all of these 'brave new world' functions will be used in non-humans for the indefinite future, and they are unlikely to appear in humans before we create *Homo nouveau* by other genetic engineering means. But what we are now seeing in genetic engineering is only the tip of the iceberg. We will move beyond somatic genetic engineering for genetic diseases to creating human enhancement, like slowing the aging process, changing physical appearance, enhancing intellectual capabilities and other traits. This will most likely be done through the genetic engineering of IVF zygotes. None of those developments require any of the changes described in this 'brave new world' section. *Homo nouveau* will not be an alien. It will, however, be a new human species.

CHAPTER 7

A HYPOTHETICAL FUTURE

"Knowledge is telling the past.
Wisdom is predicting the future."[165]

—**W. TIMOTHY GARVEY**, American endocrinologist

This chapter is complete speculation. As noted above, there is a big difference between the reporting of knowledge – as has been done in the previous chapters – and the ability to predict the future. However, as stated earlier, I believe our ability to predict the future is on firm ground.

Below I make a number of predictions based on the science described in the previous chapters. These changes, even if all come to pass, do not in themselves create *Homo nouveau*. They illustrate the advanced scientific milieu from which *Homo nouveau* will emerge. That emergence will be the result of a vast increase in genetic engineering procedures, including germline genetic engineering. We will reach the point in the next century where this can be done safely and quickly. There are a number of ways that this will lead to speciation, and this chapter will describe only one of them for illustrative purposes. It is the ubiquitous nature of future genetic engineering and the multiple ways this could induce a speciation event that gives me confidence that the next human species will be created far sooner than would be possible through normal evolution. The hypotheses I lay out in the Preface will come true and *Homo nouveau* will be created.

First, the general predictions:

- Genetic engineering will cure many cancers and single-gene-based genetic disorders, and these procedures will dramatically increase over the coming decades.
- Quantum computing will become a reality this century and will extend Moore's Law well into the future.[166] It will have widespread general use and especially facilitate greater use of AI in genetic engineering.
- One impact of more powerful AI will be on what today are called Genome Wide Association Studies (GWAS). These look for gene variants and epigenomic factors that correlate with various traits and clinical disorders. Most of our diseases, especially those in later life, are not caused by a single gene mutation. Heart disease, stroke, diabetes, Alzheimer's, arthritis and simply aging are the result of complicated genetic and environmental factors. For example, there are 270 genes associated with schizophrenia. There are more than 400 genes associated with Type II diabetes and in excess of 600 related to height. Using improved AI and databases of billions of human genomes, the genetic determinants of chronic diseases and human traits will be more precisely known and quantified.[167]
- *Polygenic scores* are mathematical calculations of the probability of various diseases or traits based on multiple genes. From the GWAS studies, it is possible to calculate the probability of developing a particular phenotype (trait or disorder), given any particular genotype, and then statistically determine the contributions of many such probabilities to an overall score to predict that phenotype. These polygenic calculations can then be applied to an individual based on his or her genome analysis. Polygenic scoring will be used for both medical interventions during routine clinical care and

pre-implantation genetic diagnosis (PGD) for embryo selection. For example, these scores could predict the probability of a person having a heart attack. This prediction would then supplement our current risk indicators, such as cholesterol levels, smoking history, exercise and dietary factors. Those indicators and polygenic scores will improve our interventions with drugs, diet, exercise and other means. Ultimately, they will guide genetic engineering of multiple gene traits and diseases during IVF.

- Advanced AI will enable us to predict most cancers at their earliest stages, using a simple blood test. Billions of microscopic fragments of DNA, RNA and proteins shed by normal and cancer cells all float around in our bloodstream.[168] With enhanced laboratory techniques and advanced AI, we will be able to detect cancer cells and identify their type to enable early intervention.[169]
- Current research has identified numerous genetic processes related to aging. For example, at the end of every chromosome in every cell there are multiple *telomeres*. These are short, repetitive DNA sequences. Every time a cell divides, the chromosome loses one telomere. After a certain number of cell divisions, there are none left, which causes that cell to deteriorate and die. That and other genetic aging processes lead to many chronic manifestations of aging, such as arthritis and loss of skin elasticity. There is intense research under way on ways to genetically alter telomeres and other aging-related genetic factors to delay or even reverse the aging process.[170]

Advances in AI and genetic engineering will not happen in a vacuum. Other sciences, environmental factors, global economics, international politics, global health and even pandemics will all interact to create our future world.

Any of these factors could be the spark – intentionally or inadvertently – leading to *Homo nouveau*.

In this chapter, I will describe one possible path to *Homo nouveau*. There are other examples I could have chosen.

THE REMAINDER OF THE 21ST CENTURY

The main environmental factor that will dominate this century is global warming. Unfortunately, the lack of international cooperation will cause the failure of most goals to reduce carbon emissions. By late century, global temperatures will have risen over three degrees Celsius. This will be due mostly to the added release of carbon dioxide and methane from melting permafrost and the loss of reflected sunlight from reduced Arctic ice and snow. Coastal cities worldwide will be flooded, causing populations to move inland. Many farmlands will become deserts and oceans will become more acidified. Food supplies will be reduced, and starvation will become greater in developing countries. Migration between countries will be curtailed because of xenophobia against refugees fleeing distressed areas.

AI and genetic engineering will advance in medicine, and both technologies will also be used to reduce global warming (examples below). AI will drive the increasing capability of genetic engineering. I predict that China will be the leader in both technologies. Here are some possible ways genetic engineering will be used:

- A significant negative consequence of global warming will be decreased rainfall in many areas, resulting in prolonged droughts. More plants and food crops will be genetically engineered to be drought resistant. New varieties of rice will not require fields to be flooded.

Resistance to GMOs will fade, particularly in developing countries where the need is the greatest.

- Livestock are major contributors to atmospheric methane, a more potent greenhouse gas than carbon dioxide. Genetically engineering livestock and their intestinal microbiomes will reduce their methane flatulence and lower atmospheric methane levels. There will also be a worldwide trend to reduce consumption of beef.
- Photosynthesis is a process that reduces carbon dioxide and increases oxygen in the air. Certain species of trees are known to be more efficient in this process than others. Using AI and GWAS, the genes responsible for this increased efficiency will be identified. Then, a massive undertaking will genetically alter the most prolific tree species in every environment and worldwide reforestation efforts will follow.
- Research in synthetic biology will create unique bacteria and other organisms that could solve specific global warming problems. For example, incorporating metallic ions onto the surfaces of algae would make them reflect more sunlight. These algae could then be deployed in the Arctic to make up for the loss of snow and ice reflections.
- Genetically engineered cyanobacteria (bacteria with photosynthetic capability) will be seeded in all of the oceans to grow and proliferate. These organisms will reduce carbon dioxide in the atmosphere while also serving as a food source for fish and other sea animals.

Undoubtedly, there will be many more ideas applied to our many problems. Not all will be successful. Even the successful implementations will take decades for their combined effects to even reduce the *rate* of global warming, let alone reducing it altogether. Global warming will be a primary catalyst for accelerating research in both AI and genetic engineering.

CHINESE TAKE THE LEAD

Over the course of this century, I hypothesize that IVF will become the preferred method of having children, particularly in China. This has nothing to do with infertility and everything to do with PGD. In 2020, China's life expectancy ranked 101st in the world. Cardiovascular disease, hypertension, chronic respiratory problems and cancer were the leading causes of death. Scientists in China will create the most advanced and accurate polygenic scores in the world for these diseases. This will be due to the combination of their massive genomic/phenotypic databases and their employment of quantum computer-based AI. So, a typical Chinese couple using IVF will ensure that their offspring have no serious single-gene disorders, and they will also choose embryos that reduce their children's chances of having chronic aging diseases. The prospective mothers will even avoid the discomforts of having eggs removed from their ovaries because of the ability to generate their eggs from stem cells. PGD will simply make IVF too attractive an option to refuse. Why would any couple not want to prevent inheritable diseases in their children if it is safe, cheap and easy to do? Generating many eggs from stem cells will have the added benefit of generating large numbers of prospective embryos from which to choose for implantation. This will provide China with the availability of large numbers of zygotes that are not chosen for implantation, which will fuel research for even more genetic engineering tools.

The most significant change in IVF will be the cautious introduction of germline genetic engineering of zygotes planned for implantation. Ironically, the He Jiankui scandal described earlier could have contributed positively to that issue. Assuming the three children he helped produce have had no genetic complications by mid-century, it could tip the scales toward allowing genetic engineering to prevent AIDS.

AIDS is a serious epidemic worldwide, including in China, and treatment of HIV infection is a costly government expense. If research on the mutant form of the CCR5 gene – the one engineered by Jiankui – continues to demonstrate HIV resistance in multiple long-term studies, this could lead to the first approved use of germline genetic engineering during IVF.

Germline genetic engineering could also be used to prevent a single-gene-related genetic disorder in children when both parents have a recessive disease, such as sickle cell anemia or cystic fibrosis. The main point is that the research with IVF embryos will have been ongoing for at least 50 years and could convince regulators that genetic engineering of such embryos can be performed safely. At some point in this century, the confidence will be sufficiently high to warrant clinical trials for IVF genetic engineering somewhere in the world.

POPULATION CONTROL

By the end of this century, global warming will subject China's five largest cities,[171] all along or near the Pacific coast, to increasing flooding and permanent land mass shrinkage. This will cause considerable migration inland and will result in overcrowding and housing shortages in some areas. Because of this migration, increasing population growth, and shrinking land areas for habitation and agriculture, China will then decide to reinstitute its one child per family policy. Between 1979 and 2015, when China executed this program, the birthrate did fall, but there were considerable problems in monitoring and enforcement. There were also unintended consequences, such as high abortion rates for female fetuses and a subsequently high ratio of men to women in the population. This time it will be much easier to control.

By early in the 22nd century, IVF will have increased to more than 90% of all live births in China. This will be due to the expansion of PGD into many non-disease traits, such as height and intelligence, which will be important for many Chinese couples in zygote selection. The government will be able to enforce its one-child policy by centralized control over the IVF clinics, and will provide an added financial incentive to couples choosing IVF. Couples will then be able to choose how their one precious child would likely turn out. It will be a win-win for all involved.

THE CREATION OF HOMO NOUVEAU

Early in the 22nd century, the world will begin to see the impact of attempts to control global warming. At first, these efforts will slow the rate of temperature increase, and then be followed decades later by actual temperature reductions. By that time, I am speculating that the seeds for *Homo nouveau's* creation will have been introduced. What follows is one way I can imagine that happening. It is not the only way, and not necessarily the most likely way, but it illustrates the general mechanism by which speciation will result from the use of genetic engineering.

By the end of this century, people all over the world will become increasingly comfortable with the safety of germline genetic engineering. The Chinese people will be the most advanced in this regard and will have expanded it to address multiple conditions, such as enabling HIV resistance and the prevention of certain recessive diseases in children when both parents have the disease. Also, China's attempts at population control will be succeeding, through its renewed one-child policy. All of this will be made possible because of genetic engineering of zygotes during IVF.

One problem with current IVF is that not all of the implants survive. And so, multiple implants are usually done simultaneously to increase the probability that at least one will go to full term. Since this could sometimes result in multiple births, Chinese authorities will limit any given IVF procedure to a single zygote implantation. If it fails, then the couple could undergo another attempt, which would be an inconvenient and costly process.

It has always been a mystery why there are not more spontaneous abortions or miscarriages, whether the pregnancy is natural or through IVF. Half of the genes of any embryo come from the father, so why doesn't the mother's immune system reject the embryo's foreign proteins? If the father contributed an organ for implantation into the mother, immune rejection of that organ would surely occur. Why doesn't that happen to the embryo in the uterus? Surely something must be going on to protect the fetus, but at this time the mechanism is unknown.

My speculation is that the Chinese scientists will do a study of all IVF procedures that fail. They will then perform a GWAS study on the aborted embryos and compare them to the genomes of all successful implants. Remember that the genomes of every implant will be known, since PGD would have been used in their selection. They will find a variation in the genes creating the surface proteins of the placenta, which comes in contact with the uterus during pregnancy. It is those proteins that will potentially be exposed to the mother's immune system. They will be able to determine which of those proteins were most associated with the failed implants and which were the least. They will find at least one such protein that is rarely associated with a spontaneous abortion and, therefore, will know the exact DNA sequence of the gene producing that protein. After appropriate clinical trials, they will then require genetic engineering of all IVF zygotes to have only that particular genetic sequence for that protein. This will result in a near 100% success rate for single IVF implantations.

Since there are more than 20 million live births annually in China, this genetic engineering procedure will be done hundreds of millions of times by the early- to mid-22nd century. By that time, the need for the genetic engineering procedure will decrease to near zero, since the genetic alteration will have been passed on to all future generations. That particular gene will become a regular part of the Chinese population gene pool. However, it might not yet be obvious to the Chinese in general that *Homo nouveau* will have been created. This will be described below.

THE IDENTIFICATION OF HOMO NOUVEAU

By the middle of the 22nd century, humanity will have reached a sort of steady state with climate change. The (by then) four-degree Celsius rise and resultant damage would have already occurred. Extreme climate events will be the norm and coping mechanisms will be put in place. Coastal cities will be smaller or even non-existent (i.e., under water) in many areas. Thanks to the genetic engineering initiatives, food production will be rising and relieving shortages in many parts of the world. Things will begin to cool down a bit (both literally and figuratively) and tensions between nations will ease. Migration restrictions will be reduced, and the long-suppressed peripatetic instincts of *Homo sapiens* will gradually reawaken.

With that migration, the first hints of the existence of *Homo nouveau* will appear. Miscarriages will now be occurring rarely in China. In fact, the only time they will be noted is when a pregnancy is the result of natural conception, where the ancestors of one or both parents had not undergone the surface protein genetic procedure. This will be increasingly rare in China, but it will occasionally occur.

Perhaps the first hint of something unusual will be noticed in Xinjiang province. This remote region of deserts and mountains is one of China's least densely populated areas. It will have a higher percentage of natural pregnancies (i.e., conception the old-fashioned way) than other provinces because of the shortage of IVF practitioners. It will be observed that the rate of miscarriages will be higher when one of the parents had the genetic procedure than when neither of them had it. That is, miscarriages will be significantly higher than previously, before any genetic engineering was being done during IVF. That will be counterintuitive, since the procedure reduces miscarriages. If anything, the incidence of those should be *lower* in these circumstances.

With increased Chinese emigration to other countries, it will soon become clear that the miscarriage rate will be near universal when a Chinese person mates with a person from a different country. It will perhaps be a little less so in neighboring countries like Korea or Vietnam, since native Chinese people will have already emigrated there in previous generations.

It will be determined that the genetic IVF procedure to alter the placental surface protein introduced a post-zygotic reproductive isolation barrier (discussed earlier in Chapter 1) into the Chinese population. The Chinese people will have become a population evolving independently from all other humans, and their genetic pool will no longer be mixing with others. They will now be *Homo nouveau*, co-existing with all other *Homo sapiens*.

CHAPTER 8

THE FOURTH GREAT TRANSFORMATION – EXPLAINED

"If we had been privileged enough to observe the origins of our species and our lineage, we would have been struck by one thing – nothing very much happened."

—**ROBERT FOLEY**, *Humans Before Humanity*

In the example in Chapter 7, the Chinese population will become a new species. They will fit the definition of a separately evolving metapopulation. I could have chosen any population for this example, and chose the Chinese people simply because of their rapidly advancing knowledge of genetic engineering. The speciation event will be due to a post-zygotic reproductive isolation barrier that prevents successful interbreeding with all other humans. That barrier is the rejection of any pregnancy involving a zygote that contains a placental surface protein that it detects as foreign. Consequently, the Chinese gene pool will not be intermixing with any other *Homo sapiens*. They will be *Homo nouveau*. The cause of the reproductive isolation barrier will be an unintended consequence of genetic engineering of IVF zygotes.

For those wishing to understand this one example of how I imagine this could happen at the molecular level, the explanation follows. Figure 10 shows the normal immune state of a woman prior to a pregnancy. It is presumed that there are multiple genes producing antibodies to foreign proteins. The strength of that antibody response varies depending on the degree of difference between the maternal and foreign proteins.

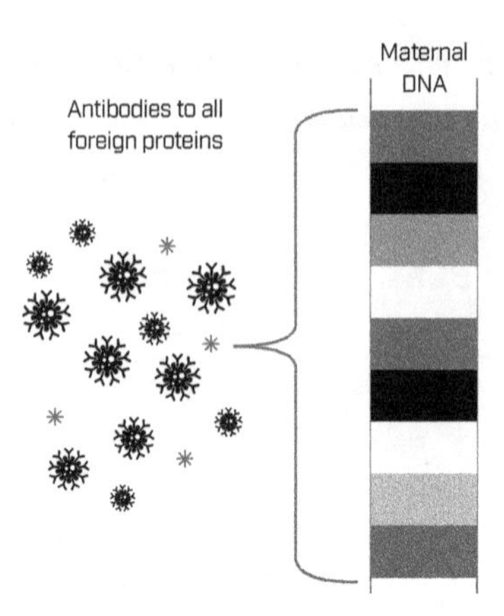

FIGURE 10 - **NON-PREGNANT STATE**

Note that the size of the antibody is meant to represent the strength of the antibody response, which varies depending upon how different the foreign protein is from the maternal version of the same protein.

For this discussion, the genes that produce placental surface proteins will be labeled as PLA genes (for placenta). In China, the most common PLA gene variants will be labeled PLA-CH. All non-Chinese variants will be collectively labeled PLA-NC. Even within a population, the PLA gene is assumed to have slight variations in its exact nucleotide sequence, but these all produce a similar surface protein. Since the fetus has one copy of the placental surface protein gene from each parent, there will be copies of the surface proteins in the placenta from each gene. This is similar to the surface proteins of red blood cells, determined by the ABO blood type set of genes. If a person with blood Type A has a child with a person with blood Type B, the child could have blood Type AB, meaning the child's red blood cells would have both types of surface proteins. Likewise, my speculation is that the placenta has a mix of surface proteins from each parental gene.

Normally, the mother carrying the fetus will produce antibodies to placental surface proteins that her immune system detects as foreign. However, evolution must have provided a mechanism for the fetus to prevent these antibodies from causing a spontaneous abortion or miscarriage. Otherwise, the species could not have survived. It is speculated here that the mechanism involves a gene or genes in the fetus that produce a substance that inhibits these antibodies. It is also assumed that the strength of the maternal antibodies is in direct relationship to how foreign the surface protein actually is, compared to the corresponding genes of the mother.

Based on the GWAS study that was done, one particular surface protein variant, whose gene will be labeled PLA-CHx, was found to be rarely associated with miscarriages. This gene itself is a very rare variant in the Chinese population, which is why it took an extensive GWAS study to determine its impact. The PLA-CHx surface protein is presumed to generate few or no antibodies from Chinese women during pregnancy, whereas PLA-CH surface proteins may occasionally produce some antibodies, since the proteins may be slightly different. However, the vast majority of fetuses with PLA-CH surface proteins go to full term. Finally, PLA-NC surface proteins would generate the strongest antibody response. Therefore, the ability of the fetus to inhibit these antibodies by its 'protective' gene is strongest in the PLA-NC situation, which would allow interbreeding of Chinese to non-Chinese people. Figure 11 illustrates the normal situation in China during pregnancy, when the mother is Chinese.

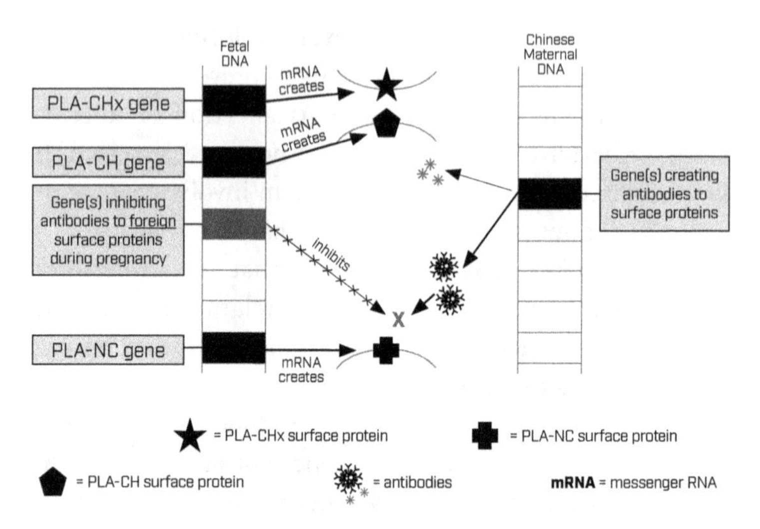

FIGURE 11 – **NATURAL STATE DURING PREGNANCY**

In this figure, examples of three possible surface proteins are shown. In any given pregnancy, only one or two of them would likely be present. The figure simply illustrates the possible interactions of the immune systems between the mother and the fetus. If the mother is non-Chinese, mating with a Chinese father, the strong antibodies would be directed against the PLA-CH or PLA-CHx surface proteins (rather than the PLA-NC surface proteins), yet still be inhibited because of the protective gene in the fetus.

Once the Chinese scientists determined that the mother produces no antibodies to the PLA-CHx surface protein, they then required that all future IVF implants be modified to carry only that form of the gene. This genetically modified version of this gene will be labeled PLA-CHxge. All future generations would inherit that gene as well. Over time, the need to perform that genetic modification would reduce to near zero, since nearly all Chinese people would have the engineered PLA-CHxge gene.

It is speculated here that the genetic engineering of the PLA-CH gene to be identical with the naturally occurring PLA-CHx gene had an unintended off-target mutation, in addition to its intended correction. That off-target mutation created an inhibitor to the expression of the fetal gene providing the protection against the maternal antibodies. The most likely mechanism was the creation of a specific RNAi molecule, which is a common natural mechanism used to inhibit gene expression.[172] The newly engineered PLA-CHxge gene produces the identical surface protein as the naturally occurring PLA-CHx variant, but the gene differs from that variant by creating the unwanted RNAi.

So, when a mother with the PLA-CHxge gene has an IVF implant from a father with the PLA-CHxge gene (or the natural PLA-CHx gene), the embryo would almost always carry to a normal full term, since the mother produces no antibodies to the PLA-CHx surface protein. That, of course, was the intent. If the father had an unmodified PLA-CH gene, spontaneous abortions would still occur as before, but not in a high number. Most of those babies would also go to full term, as they did before the genetic engineering began, but at a statistically lower rate because there would still be some minor uninhibited antibodies. Eventually, there were very few Chinese people without the PLA-CHxge-produced surface protein. If the fetus had a PLA-NC surface protein, a miscarriage would always occur, because of the now uninhibited and strong maternal antibodies to that surface protein. Figure 12 shows the *Homo nouveau* state during pregnancy when the mother is Chinese.

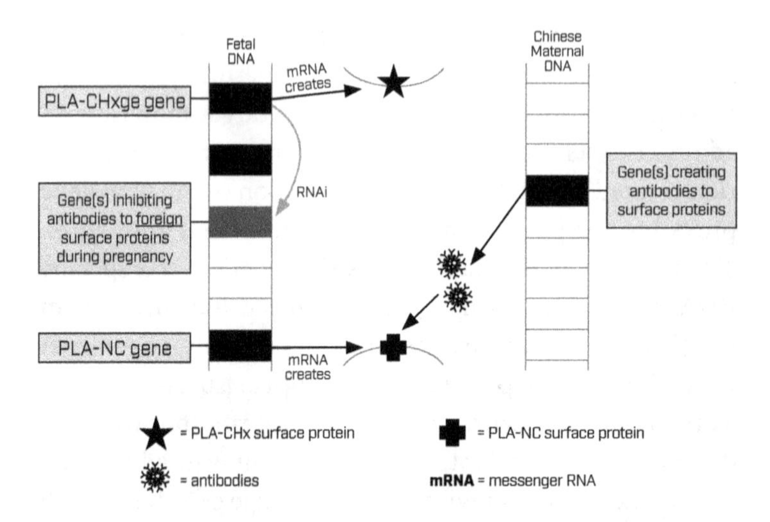

FIGURE 12 – ***HOMO NOUVEAU* DURING PREGNANCY – CHINESE MOTHER**

If the father is also Chinese, he would likely have the PLA-CHxge gene and the fetus would almost always carry to full term. In the case where a Chinese woman emigrates to another country and the father is non-Chinese, the uninhibited antibodies to the PLA-NC proteins would always lead to a miscarriage.

Figure 13 shows the *Homo nouveau* state during pregnancy when the mother is non-Chinese, which would be the likely case when a man emigrates to another country.

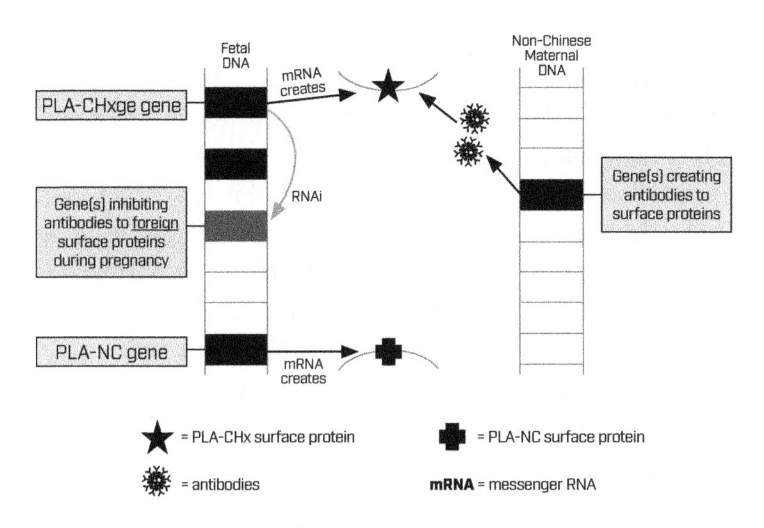

FIGURE 13 – ***HOMO NOUVEAU* DURING PREGNANCY – NON-CHINESE MOTHER**

In this case, the uninhibited maternal antibodies would be directed against the PLA-CHx placental surface proteins from the father, causing a miscarriage.

Why then did they find a higher rate of miscarriages in Xinjiang province? It turns out that there were more natural pregnancies in this province due to the lower availability of IVF practitioners. In addition, there is a higher percentage of people of non-Chinese origin in that autonomous border region. Thus, there were more pregnancies with placental surface proteins of PLA-CH and, more importantly, of PLA-NC.

When Chinese people, whether male or female, began to emigrate to other countries, they now contained the PLA-CHxge gene as well as the inhibitor to the protective gene. When they interbred with any non-Chinese person, the fetus would always have foreign placental surface proteins that were unprotected from maternal antibodies and miscarriages would occur. As noted, this happened slightly less often in neighboring Korea and Vietnam because of the presence of Chinese people living in those countries.

Keep in mind that people with the PLA-CHxge gene not only create the PLA-CHx surface protein. They also are unable to inhibit the maternal antibodies to foreign proteins. This is why non-Chinese mothers in foreign countries with PLA-NC genes would have near universal miscarriages with male Chinese fathers with the PLA-CHxge gene, while in the past they did not. However, they could still have a normal offspring with a Chinese person with a natural (i.e., non-genetically engineered) PLA-CHx or PLA-CH variant. Likewise, if a Chinese woman with the PLA-CHxge gene emigrates and mates with a non-Chinese male, their fetus would also miscarry. Table 5 summarizes the miscarriage frequency with all of the various genetic PLA genotype combinations. The first column lists all of the combinations that the two parents could have of the placental surface protein genes. It does not matter which parent has which gene.

FETUS GENOTYPE	MISCARRIAGE FREQUENCY
PLA-CHxge / PLA-CHxge	Rare
PLA-CHxge / PLA-CHx	Rare
PLA-CHxge / PLA-CH	Infrequent
PLA-CHxge / PLA-NC	Always
PLA-CHx / PLA-CHx	Rare
PLA-CHx / PLA-CH	Infrequent
PLA-CHx / PLA-NC	Infrequent
PLA-CH / PLA-CH	Infrequent
PLA-CH / PLA-NC	Infrequent
PLA-NC / PLA-NC	Infrequent

TABLE 5 – **MISCARRIAGE FREQUENCIES BY FETUS GENOTYPE**

I know this is complicated, but that is always the case with the genome and genetic engineering. That's why we need AI. This speculative example relates to our current lack of understanding about why the mother doesn't always reject the father's foreign proteins in the developing fetus. We know that *something* must be going on at the molecular level that prevents this. Our research does not explain completely the mother's immune response or the fetus' response to it.[173,174] The Chapter 7 example could happen to any population undergoing a common genetic engineering procedure. Note that the example also illustrates the dangers of germline genetic engineering. An unintended off-target mutation could occur in addition to the successful intended one.

In the Chapter 7 example, the IVF genetic engineering procedure introduced one of the necessary ingredients to speciation: a reproductive isolation barrier. Such barriers are what allow new species to develop and evolve away from their predecessor species. As discussed in Chapter 1, there are many types of reproductive isolation barriers. What is unique in this example is that for all of human history, and for all of the human species that ever lived, this would be the first time that a post-zygotic barrier enabled the emergence of a new human species.

All previous human speciation events had physical separation of small groups as their pre-zygotic isolation barrier. That is, the reason that a new human species emerged in the past was that a group had become physically separated by migration from their predecessor species for thousands or even millions of years. During this period, they evolved independently to the point where both the genetic and physical differences became sufficiently large for speciation. Historically, all of the human speciation events have been due to pre-zygotic reproductive isolation barriers.

The other necessary requirement for a new species, then, is that it is evolving independently from all other species.

Remember from the discussion of races in Chapter 2 that our current racial groups are not evolving independently from other *Homo sapiens*. There is free mixing and interbreeding of people of all racial and ethnic backgrounds, which causes constant remixing of the genetic components of *Homo sapiens*. That will not occur with *Homo nouveau*.

In the case of the genetically engineered Chinese people, it would be likely that even after a hundred years they would all still look the same, speak the same and be culturally the same as previous Chinese people. During the transition period, which would happen in place without any migration, they would be visibly indistinguishable from those Chinese people that had not been genetically engineered.

How could they then be called a separately evolving metapopulation, as the definition requires? The answer is that even if the transition took a hundred years, that is far too short a time to have any obvious physical evidence of evolutionary change. That is precisely why I started this chapter with the quote from Robert Foley in *Humans Before Humanity*.[175] Had we been around when *Homo sapiens* first emerged from *Homo heidelbergensis* (or whomever else we emerged from), we wouldn't have noticed any difference either, for perhaps a thousand or more years.

However, unlike racial groups today, we could certainly identify at least one unique set of genetic features of all *Homo nouveau* that 100% of them would have. That would be the PLA-CHxge gene in the example. Although some *Homo sapiens* would also have the naturally occurring PLA-CHx gene, certainly not all of them would, and none would have the PLA-CHxge gene. If we looked hard enough after a hundred years, we might even find other consistent genetic differences between *Homo nouveau* and *Homo sapiens*. Certainly, other differences will evolve over millennia, since the genomes of *Homo nouveau* will be evolving independently of *Homo sapiens*.

EPILOGUE

This book is entirely non-fiction except for Chapter 7, which *is* fiction, and Chapter 8, which explains the genetics and molecular biology of Chapter 7. These last two chapters are my best attempt to describe a potential, but realistic, future scenario based on the science described in Chapters 1–6. My intent is not to create a science fiction story, but rather to make the real science – and a particular outcome – as vivid and realistic as possible. Chapters 7 and 8 are meant to be an educational illustration of how *Homo sapiens* might create *Homo nouveau* over the next century.

I have described in detail one of many possible scenarios that could create a new human species by using genetic engineering augmented by AI. The *Fourth Great Transformation* won't necessarily map to that particular scenario, which is just one example of how speciation can occur at the hands of humans in the future.

Up until now, all speciation of all species, including humans, has occurred by various naturally occurring mechanisms. From now on, my hypothesis is that the speciation of the *Homo* genus will no longer occur in that manner. Instead, it will be caused by whatever

Homo species exist at any given time, using genetic engineering. That will be the true *Fourth Great Transformation.*

For more than 99% of the time that humans have existed, there were multiple human species co-existing on Earth. It is only during this latest period of about 20,000 years or so that there has been only one species of humans – *Homo sapiens*. This brief, exceptional period of a single human species will end with the arrival of *Homo nouveau*. And, it is also likely that there will be more than two human species at some point in the future, because there are multiple possible unintended results of genetic engineering that could result in other reproductive isolation barriers.

The hypothetical example that I discussed was an attempt by Chinese authorities to reduce miscarriages by performing millions of IVF procedures for the same genetic alteration over a relatively short period of time. Other examples that could have been chosen are the development of a genetic prevention to a common disorder like Alzheimer's disease, a method to reduce or reverse the aging process, a possible genetic engineering approach to a pandemic, a treatment for obesity[176] or any number of possible human enhancements that millions of people would want to undergo for themselves or for their future children. It could happen to any group in the world having access to genetic engineering. It does not need to occur in groups as large as the Chinese population. In fact, one could argue that it would be more likely if I had chosen a smaller or more isolated population example.

Further, the off-target mutation doesn't need to relate to immunity during pregnancy nor even to the creation of a post-zygotic barrier. For example, there is a publication on a National Institutes of Health website reviewing the genetics of sialic acids, which are ubiquitous substances on the surfaces of various cells. A mutation in a single gene differentiates humans from other animals for many of these substances. Such a mutation, if occurring on genes related to sperm, could render sperm unable to ever reach an egg for fertilization in the woman's uterus. The article concluded that "this could even be a mechanism of speciation in the genus *Homo...*"[177] This would be an example of a pre-zygotic barrier, but a different kind of pre-zygotic barrier than the one that lead to *Homo sapiens*.

The main point is that there are multiple ways that reproductive isolation barriers can be introduced by genetic engineering. My hypothesis is that such barriers will be unintended and unnoticed for many generations. They will not be deliberate attempts to create a new species.

This hypothesis does *not* imply that evolution by Darwinian natural selection has ended in humans. There is much evidence of natural selection in the most recent 10,000 years. It is why most Europeans are now lactose tolerant as adults when 10,000 years ago they were lactose intolerant. The domestication of milk cows and the abundant consumption of cow's milk throughout life has led to that. It is why the first humans settling in Great Britain were dark-skinned, but over time were replaced by light-skinned humans to accommodate the less sunny environment. Australian Aboriginals developed a genetic variant in the most recent millennia to accommodate extremely high temperatures. Likewise, Tibetans can breathe easier at high altitudes and Inuits can digest fish fats better than others. Some Argentinians are more tolerant of arsenic because of the high levels in their drinking water.

Random mutations and Darwinian natural selection will continue indefinitely, having some effect or another on the evolution of *Homo sapiens* and other future human species. That effect has likely been reduced dramatically in the past 50–100 years because we are better able to keep people with harmful genes alive past reproductive age. That is, negative natural selection is less effective in preventing harmful genes from propagating. However, natural selection is still going on and will continue indefinitely.

My hypothesis is that *speciation* as a result of Darwinian natural selection has ended for humans. New human species will not evolve naturally. Within every human species – including *Homo nouveau* – genetic mixing will continue to prevent a separately evolving metapopulation from developing within it. Instead, other types of human-created barriers will be produced by genetic engineering, which will lead to other human species. Given our population size, geographic extent and mobility, there simply is no way to create the conditions that existed in the past to allow a natural isolation of a human metapopulation.

How will *Homo nouveau* and *Homo sapiens* get along? Many different factors could come into play over the next millennia to determine that. During the transition period until virtually all Chinese people have the new placental surface protein gene, those with and without that gene will be indistinguishable in every way except by genetic analysis. At first, they will surely get along, for better or worse, the same as any other groups currently get along. Memes will cross species lines and that should keep the two species culturally similar for a long time. In fact, it is unlikely that the *Homo nouveau* people would refer to themselves as a different species or even a different race. Nor would the media or other institutions differentiate them. *Homo nouveau* would simply be another population co-existing with everyone else as our populations do today. Except, of course, their genes won't remix and they will evolve separately.

Many influences could alter one or both species in significant ways over time. One possibility would be some type of infection that affects one species more than the other. Or, perhaps, they could react differently to certain environmental factors that cause cancer or other diseases. In the past, there have been many examples of related species of the same genus co-existing in the same areas that have gone extinct at different times, including humans. This seems less likely to me for human species going forward, however, given our current tools and the new tools undoubtedly yet to come.

It is also possible that once the genetic differences are understood, there could be some massive attempt to reverse those differences by genetic engineering. That would also be unlikely. First, it is not clear if there would be motivation enough to invest in such an expensive effort. Second, not everyone would wish to participate. For those mixed-species couples desiring to have normal children, they could always use IVF and genetically engineer a correction to the PLA gene or the off-target mutation on a one-off basis.

Speciation happens over thousands, or millions, of years. There has not been a speciation event in our lineage for 300,000 years. My speculated example will occur within a couple of centuries – a blink of the evolutionary eye. That is a dramatic change that will likely be repeated many times.

I have not mentioned the singularity in these final chapters. I don't think any form of it will happen in my speculated time period. In particular, with regard to the Kurzweil singularity, I don't believe that nanobots will ever be able to download the contents of an individual human brain into a computer such that that person will exist in any definable way electronically. I'm even more skeptical that nanobots will be able to upload new capabilities into our brains. I do think that quantum computers will outperform humans in virtually any task they are assigned to do. I'm not concerned, however, that they will be an existential threat to humans. Although unintended consequences will occur, I believe we will be able to intervene to mitigate them.

There will be time for a *Fifth Great Transformation*.

ACKNOWLEDGMENTS

IA – INTELLIGENCE AUGMENTATION

As its title implies, AI plays a major role in this book. It is one of the two enabling tools, along with genetic engineering, in the creation of *Homo nouveau*. As I've pointed out repeatedly, the vast majority of AI today does not replace human intelligence but rather augments it. It is IA.

IA is what enabled me to write this book. I'm a physician. I've had extensive science and medical training. My medical career focused heavily on informatics – using computers for medical decision support and electronic medical records. That enabled the AI focus. I'm also a researcher, having published many peer-reviewed studies in my field. But I am not an evolutionary biologist, nor a taxonomist, nor a geneticist, nor an expert in the many fields that I needed to review and understand in order to credibly write this book. Fortunately, my background in medicine, science, informatics and research has given me the brain tools to be able to read the science literature in these various fields and understand it sufficiently to bring it together. It is the intersection of all of these fields of study that is necessary to allow an informed speculation on the future speciation of *Homo sapiens*. There is no expert on that subject. Much of my basic understanding came from the many full-length books on the various areas referenced here. But things are changing rapidly in many of these fields. Therefore, I needed to augment my basic understanding using one of the most powerful tools for information transfer ever created by *Homo sapiens*: the internet.

Insight into virtually all of today's science developments is available one way or another through the internet. Most of the time I needed to access the original publications in *Science, Nature, Proceedings of the National Academy of Sciences, New England Journal of Medicine, JAMA* or more specialized journals, such as *Systematic Biology, Nature Nanotechnology, Minds and Machines* and many others. Google Scholar was a particularly useful source for these original articles. Since many of these science fields, particularly genetic engineering, are moving so rapidly, much of my material came from pre-publication sources such as press releases and pre-peer-review online journals, as well as many direct, personal communications with scientists and researchers.

The latter communications were particularly useful and appreciated. Since I am not an expert in all of these fields, I only claim to be able to understand them to a level necessary to ask the right questions when I needed clarification. Many of the real experts in their fields were kind enough to take my phone calls or emails when I needed such clarification. I am particularly grateful to Kevin de Queiroz, Svante Pååbo, Ann Gibbons, Eugene Harris, Aubrey de Grey, David Reich, Katherine Pollard, Chris Stringer, Nessa Carey, C. Owen Lovejoy and Ian Tattersall in that regard. Each of them was gracious enough to respond to my specific inquiries, to enable me to clear up details when I needed that help.

EDITORS

Getting the facts right was necessary, but not sufficient, to write this book. I needed to write it in a manner that could be understood – and enjoyed – by the broader reading public. That turned out to be a much bigger problem. This book went through 25 versions over three years before this published version. Some of the revisions were due to the fact that the science, particularly in genomics and genetic engineering, was changing so rapidly that the earlier versions became out of date. The bigger problem was to make it more understandable and to find the right level of technical detail to avoid turning it into a series of graduate-level science lectures.

Peter Beren, my literary agent, deserves most of the credit for shaping my voice. After initially declining to be my agent after reading an early manuscript of my first book, *What Comes After Homo Sapiens?*, he agreed to have lunch with me just to give me some advice on how to go forward. We met at the historic Hotel Mac in Point Richmond, CA, near his home. The Hotel Mac is a special place. In addition to serving excellent food, there is an area near the bar consisting of unmatched old sofas, lounge chairs, lamps and coffee tables, reminding me of furnishings one might get at a rummage sale. They made us a perfectly comfortable place to spread out with a computer and papers and just talk quietly while lunching on the local clam chowder. Sitting there at our first lunch together, I shared with Peter my enthusiasm for human evolutionary biology and the impact on it that modern technology might have. After an hour or so, Peter said to me, "If you would write the way you talk, I'd be much more interested in your book." At that point, he agreed to be my writing coach as well as my agent. For the next year or so, Peter and I met regularly at the Hotel Mac to review sections of my written work and talk mostly about style and voice. This book is one of the results of these Hotel Mac visits. The more important result is my friendship with and admiration for Peter.

Although Peter was my agent and writing coach, he was not my editor. Since my goal was to make the book accessible to readers of non-fiction who aren't steeped in science backgrounds, I tried out earlier versions on friends and relatives before submitting it for editorial review. As usual, my wife Madeleine, a family law private judge and an avid book reader with little background or interest in science, took the first crack at a chapter or two. That was my first warning that I had initially missed the mark. During those early drafts, I was also involved in founding a new company aimed at using a digital medicine approach to the opioid crisis. (That's for another book.) One of the company co-founders, Brian Perkins, is an accomplished pharmaceutical company executive whom I first met when starting the enterprise. In addition to our business relationship, he and I became good friends. When he learned I was writing a book, he graciously offered to be an early reviewer. Fortunately, that did not end our friendship, but it was my second warning that the book was not ready for the lay reader. He was gentle but clear: it needed extensive editing.

By about the 15th version of the book, the COVID-19 pandemic had confined my wife and me to sheltering in place, which gave me lots of time to work on the book. Our only real outlet was to get out and hike among the beautiful San Francisco Bay area trails. One of our hiking buddies happened to be another accomplished literary agent, Amy Rennert. On hearing of my need for an editor, she connected me to the first of two major editors I used prior to submitting the manuscript to the publisher. His name is Rick Clogher. Rick is a long-time professional editor of both fiction and non-fiction works, whom Amy said could take a crack at this. In retrospect, it was more than a crack.

I'd say more like a colonoscopy or root canal work. I wouldn't call it fun, but it was the medicine I needed to revise and cut major sections of the book, including eliminating entire chapters and appendices. He was right on, and I am thankful to Rick, as well as Amy for making the connection. At the end of Rick's work, he suggested that I undergo yet another round of editing with someone specifically oriented toward the science aspects of the latest version. That led me to my second major editor, Dennis Sides. Dennis is the perfect test case for the book, besides being an editor. He is not a scientist, but rather an avid lay reader of science non-fiction books. That's just the type of editor I was looking for. If he doesn't like it, I figured, no one will. If I use the analogy that Rick Clogher was the surgeon, then Dennis was my physical rehab person. His thoughtful rewording of sentences and even whole paragraphs brought clarity to my complex prose. In addition to these structural edits, he raised questions that other thoughtful readers would raise, challenging me to answer them. And finally, it was Dennis who gave me the wonderfully quotable quote near the end of Chapter 5: "Perhaps our ultimate irony is that the human brain will never understand the human brain." In summary, the combination of Rick and Dennis got the manuscript ready for submission to LID Publishing.

They, then, took it from there. From book cover design, to further structural editing, to layout, printing, distribution and marketing, they brought this edition to life. I am particularly grateful to Aiyana Curtis for her editorial management and Caroline Li for the cover design. A special thanks goes to the copy editor, Brian Doyle. Amazingly, after countless rereads by myself and detailed review by friends and three structural editors, he still managed to find and correct many typos and other errors that had been overlooked. More significantly, Brian provided meaningful rephrasing of many of my convoluted sentences.

This not only demonstrated his editing ability, but reflected obvious additional literature reading on his part. A wonderful independent graphic designer, Eleanor Johnson, provided Figures 1–3. A special thanks goes to Carol Reed of BlueStem, LLC for her masterful job of indexing the book. My now deceased literary friend, Barbara Bonfigli, insisted that I use the term *Homo nouveau* for the new species, rather than a traditional Latin name. After all, the *Fourth Great Transformation* does not follow evolutionary biology tradition.

GLOSSARY OF TECHNICAL TERMS AND ABBREVIATIONS

AGI

Artificial general intelligence – the simulation in a computer of generally all human intelligence.

AI

Artificial intelligence – the simulation in a computer of some component of human intelligence.

AIDS

Acquired immune deficiency syndrome – a disease caused by the HIV virus, causing impaired immune defenses to infections, tumors and other complications.

Alternative splicing

The process whereby a single gene can produce multiple proteins.

Amino acids

Chemical compounds that are the building blocks of proteins. Of the more than 500 types of amino acids in nature, only 20 are used naturally to build the proteins in living organisms.

Anthropology

The study of the origins, physical and cultural development, biological characteristics, and social customs and beliefs of humans.

Archaea

Single-celled organisms that constitute one of the three major domains of living organisms. The other two domains are Bacteria and Eukarya. Archaea differ from Bacteria primarily in their cell wall structures.

Archaic *Homo sapiens*
The group of species in the *Homo* genus that preceded *Homo sapiens* in the evolutionary lineage.

Ardipithecus
A genus of extinct Hominidae that lived 4–6 million years ago. They may be in the *Homo* lineage, and walked upright when on the ground, but had an opposing big toe for better tree navigation.

ASI
Artificial superintelligence – the simulation in a computer of intelligence that far exceeds human intelligence.

Australopithecus
A genus of extinct Hominidae that lived 3–4 million years ago. Lucy is the most famous fossil. Walked upright without an opposing big toe. May be the predecessor to the *Homo* genus.

Autosomal
Pertaining to the chromosomes that are not sex-linked. In humans, there are 22 pairs of autosomal chromosomes and one pair of sex-linked chromosomes.

Bipedal
Terrestrial locomotion – upright walking – using two legs.

CAR-T
Chimeric antigen receptor T-cells – these are genetically engineered immune cells created to treat certain cancers.

Cas9
CRISPR Associated Protein 9 – an enzyme used in conjunction with CRISPR in genetic engineering to cut DNA. Other enzymes, such as Cpf1, have also been discovered that serve the same purpose as Cas9.

Chloroplasts
The organelles (subcellular structures with specific functions) within plant cells in which photosynthesis occurs.

Chromosomes
The threadlike structures within the nucleus of eukaryote cells containing the genetic DNA. Humans have 23 pairs of chromosomes in each cell.

Coding gene
A gene that defines the template for a protein or set of proteins. In common usage, a gene is the same as a coding gene.

Codon

A set of three nucleotides defining the template for a specific amino acid, or the 'stop' instruction in protein synthesis directed by DNA. There are 64 codons in the genome of organisms.

Connectome

A comprehensive map of neural connections in the brain and other components of the nervous system of an organism.

CRISPR

Clustered regularly interspaced short palindromic repeats – a DNA sequence found naturally in bacteria which, in conjunction with Cas9, serves as an immune defense mechanism against viruses. Modifying the CRISPR sequence has become a powerful tool for genetic engineering.

Darwinian evolution

The process of species modification during evolution as described by Charles Darwin, in which environmental factors determine which combinations of genes provide traits (phenotype) that best lead to survival and reproductive success. This process is called natural selection.

Denisovan

A member of an extinct species of humans living in Asia at the same time as the Neanderthals, known originally by genetic analysis of a few small fossil fragments found in the Denisova Cave in Siberia.

Deoxyribonucleic acid

A string of permutations of four nucleotides that constitute the genetic material. The nucleotides are cytosine, thymine, guanine and adenine. Each sequence of three nucleotides constitutes a codon defining the inclusion of a specific amino acid into a protein.

DNA

Deoxyribonucleic acid.

Domain

The highest level of taxonomy, consisting of three domains: Archaea, Bacteria and Eukarya. All living organisms fall into one of these three domains. Domains are further divided into kingdoms, phyla, classes, orders, families, genera and species.

ECoG

Electrocortography – a technology in which electrodes on the surface of the brain record signals that are communicated to a computer and interpreted to perform various functions.

EEG

Electroencephalogram – a diagnostic test in which electrodes on the surface of the skull record brain wave signals. This non-invasive technology has been used for decades to diagnose epilepsy, sleep disorders and other brain conditions.

Epigenetics

The study of the epigenome.

Epigenome

The set of chemical processes that affect the expression of genes. These consist of various non-coding genes and chemical mechanisms, including various RNA sequences, methylation and others. The epigenome does not alter the DNA sequence.

Eukarya

One of the three domains of life characterized by species with cells containing a nucleus surrounded by a nuclear membrane. All plants and animals are in this domain. The other two domains are Bacteria and Archaea, both consisting of single-celled organisms.

Eukaryote

A member of the Eukarya domain characterized by species with cells containing a nucleus surrounded by a nuclear membrane.

Evolution

Change in the heritable traits of biological populations over successive generations by such processes as mutation, natural selection and genetic drift.

Evolutionary biology

The study of evolutionary processes.

Exaptation

In evolution, a feature or trait of an organism originally adapted or selected for one function that is later co-opted to perform another function. For example, jawbones in reptiles were later adapted to be hearing ossicles in mammals.

Expression

As related to genes, the degree to which a gene is operational in any given cell at any given time in an organism's development. Gene expression is governed by the epigenome.

Family

The taxonomic level just above genus that usually includes multiple genera.

Fertilization
The union of a male and female gamete. In humans, this is the union of a sperm and egg to form a zygote.

Fictive thinking
The ability to communicate (fictive language) and think about (fictive thinking) fictitious or imaginary concepts and about concepts that have no objective reality, such as countries, corporations and organizations. It is a defining characteristic of *Homo sapiens*.

fMRI
Functional magnetic resonance imaging – a technology that allows visualization of blood flow in the brain that reflects brain activity. It is useful in mapping brain function as well as brain structure.

Fossil
The preserved remains or imprints of organisms from previous geologic eras.

Gamete
A mature male or female reproductive cell that combines with a gamete of the opposite sex to form a zygote. A gamete has only one of each chromosome type (haploid) rather than a pair of each chromosome type (diploid). The combination of two gametes during fertilization produces a complete diploid DNA set.

Gene
The basic unit of heredity, consisting of a DNA sequence. General usage implies a coding gene that defines the template for a protein or set of proteins. There are also non-coding genes.

Gene drive
A genetic engineering technique that leads to a particular gene variant having greater than the normal 50% chance of being inherited.

Gene knock-in
In genetic engineering, the ability to add a normal gene segment or an entire normal gene to replace an abnormal gene segment or abnormal gene.

Gene knockout
In genetic engineering, the ability to disable, replace or otherwise render inoperable a gene producing a deleterious effect.

Gene pool
The set of all genes in the population of a particular species.

Genetic drift

The random change in the frequency of gene variants in a population over time.

Genetic engineering

The direct manipulation of an organism's genome using biotechnology.

Genetics

The study of genes, genetic variation and heredity in living organisms.

Genome

The complete set of inheritable genetic material of an organism. This includes genes, epigenomic DNA, other DNA and, in RNA viruses, the RNA genetic material.

Genomics

The subset of genetics that focuses on the study of the structure and function of the genomes of organisms. DNA sequencing, data analytics, epigenomic mechanisms and studies of 3-D structure are all components of genomics.

Genotype

The set of DNA sequences that determine a cell or organism trait.

Genus

The taxonomic level just above species. A species is defined by the combination of its genus and species names.

Germline cells

Cells whose genetic makeup is passed on to the next generation. These include embryonic stem reproductive cells, gametes (egg and sperm in humans) and zygotes.

Heterozygous

Having two different variants for the same gene.

HIV

Human immunodeficiency virus – the virus that causes AIDS.

Hominid

Any of the living or extinct primates of the family Hominidae, including all species of great apes and the genera *Homo*, *Australopithecus* and *Ardipithecus*.

Hominin

Any of the species of humans and their immediate ancestors, including *Australopithecus and Ardipithecus*. Hominin does not include great apes.

Hominidae

A taxonomic family of genera that includes extant great apes and humans, and multiple extinct genera, including extinct *Homo species*, Ardipithecus and Australopithecus

Homo erectus

An extinct *Homo* species that lived from 1.8 million years ago to about 100,000 years ago – the longest surviving human species. Probably in the direct *Homo sapiens* lineage.

Homo heidelbergensis

An extinct *Homo* species that lived 600,000–200,000 years ago, and probably the direct ancestor to *Homo sapiens*, Neanderthals and Denisovans.

Homo nouveau

Imaginary future *Homo* species.

Homo sapiens

Today's modern humans. Emerged in Africa about 300,000 years ago and spread throughout the world.

Homozygous

Having both copies being the same for a specific gene.

IA

Intelligence augmentation. This is distinguished from AI (artificial intelligence) in that AI usually replaces intelligent functions, whereas IA supplements or augments intelligent functions.

IVF

In vitro fertilization.

Kingdom

In taxonomy, the classification just below the highest level (domain). Humans and all other animals are in the kingdom Animalia.

Lamarckian evolution

The philosophy promulgated by Jean-Baptiste Lamarck, a French biologist, in which traits are inherited based on excessive use or disuse of functions. Also, offspring manifest a blending of parental traits that were acquired during a lifetime. This theory was largely debunked and replaced by the concepts of Mendelian inheritance. We now know that some epigenetic changes to the genome, as well as germline cell mutations, which are acquired during a lifetime, are passed on to offspring.

Lineage

A sequence of species, each of which is considered to have evolved from its predecessor.

Mass extinction

The extinction of more than 50% of living species in a geologically short period of time. There have been five such mass extinctions in the history of Earth, and some believe we are currently undergoing the sixth.

Meiosis

In sexually reproducing organisms, the process whereby specialized embryonic stem cells (in the testis and ovary of humans) go through special cell divisions to produce the gametes (sperm and eggs). In the process, the fully paired complement of chromosomes (diploid) is reduced to half (haploid) in each gamete. Each haploid chromosome contains a mixture of genes from each parent.

Mendelian genetics

The laws of inheritance as first described by Gregor Mendel, in which discreet inheritance units (later named genes) can be dominant or recessive and result in predictable ratios of offspring traits (later called phenotype) depending on the homozygous or heterozygous nature of the genes (later called the genotype). Traits are inherited as 'all or none' rather than blended.

Messenger RNA

A large family of RNA molecules that convey genetic information from DNA to the ribosome, where they specify the amino acid sequence of a protein. Usually labeled mRNA.

Metadata

Data about other data. This often refers to labels explaining images or processes.

Metapopulation

A subset of the total population of organisms sharing a common gene pool.

Methylation

When referring to DNA (DNA methylation), a process by which methyl groups are added to DNA. Methylation modifies the function of the DNA. When located in a gene regulator, DNA methylation typically acts to repress gene transcription and expression, usually permanently. Methylation is an epigenetic process that does not alter the DNA sequence.

Mitochondria

Small organelles found in the cytoplasm of most eukaryote cells. They provide the power or energy for cell metabolism. They contain a small number of genes that are inherited only from the mother. In evolution, they are remnants of ancient bacteria.

Mitosis

In multicellular organisms, the process of somatic cells dividing into two identical new cells containing the full diploid complement of chromosomes. This is also how non-sexually reproducing single-celled organisms reproduce.

Molecular biology

The science of the molecular basis of biological activity between biomolecules in the various systems of a cell, including the interactions between DNA, RNA and proteins and their biosynthesis, as well as the regulation of these interactions.

MRI

Magnetic resonance imaging – a non-radiation imaging technology using powerful magnets that is useful in studying the structure of the normal and diseased brain, as well as other organs and tissues.

mRNA

Messenger RNA – the RNA produced by a coding gene that leaves the nucleus and migrates to the ribosome to participate in the production of a protein.

Mutation

A permanent alteration of the nucleotide sequence of the genome of an organism. These alterations occur spontaneously during cell replication or as the result of various factors, such as radiation and exposure to toxins.

Nanobot

A very small robot or computer device that can be injected into the blood stream or for other uses. They range in size from 1–100 nanometers (a nanometer is one billionth of a meter). Potential uses in humans include cancer treatment and brain communication to and from a computer.

Nanotechnology

Manipulation of matter at the atomic, molecular and supermolecular scale, including the manufacture of products with at least one dimension in the 1–100 nanometer scale (a nanometer is one billionth of a meter).

Natural selection

A key mechanism of evolution in which an organism's traits (phenotype) interact with the environment to create differential survival and reproduction rates. It is the key concept of Charles Darwin's theory of evolution (Darwinian evolution).

Neanderthal

A member of the extinct species *Homo neanderthalensis* that lived in Eurasia from approximately 600,000 to 39,000 years ago. They are closely related to *Homo sapiens* but not in our direct lineage. There was a small amount of interbreeding between Neanderthals and *Homo sapiens.*

Neuron

The basic cellular unit of the brain and nervous system that transmits, conducts and receives electrochemical signals from other neurons.

Neuroscience

The science of the study of the nervous system.

Neurotransmitter

A chemical substance facilitating the communication between neurons across a synapse. There are many types of neurotransmitters.

Non-coding gene

A DNA sequence that does not code for a protein but rather codes for various types of RNA that are important in the epigenome. Thus, it is a misnomer to label it as non-coding. This is distinguished from a 'coding gene' (what is usually referred to as a 'gene') that codes for a protein.

Nucleotide

An organic molecule that is the building block of DNA and RNA. DNA is composed of four nucleotides: adenine, thymine, cytosine and guanine. In RNA, uracil replaces thymine.

Paleoanthropologist

A person who studies human evolutionary development and lineages by studying fossils, indirect evidence such as tools and dwellings, and genomics.

Paleogenomics

The study of genomes of extinct species obtained from fossil extractions.

PGD

Pre-implantation genetic diagnosis – an IVF technique that allows the complete genetic analysis of a fertilized egg to determine if it is free of specific disease-causing genes or other genetic characteristics prior to being chosen for implantation in a patient.

Phenotype
The appearance or manifest trait that is determined by a DNA sequence.

Photosynthesis
The process of converting sunlight and carbon dioxide into sugar and oxygen within plant cells, algae and certain bacteria.

Postzygotic
Occurring any time after fertilization and the formation of a zygote, including the entire gestational period.

Prezygotic
Occurring prior to the fertilization and formation of a zygote.

Punctuated equilibrium
A theory in evolutionary biology, which proposes that once species appear in the fossil record they will become stable, showing little evolutionary change for most of their geological history. This is in contrast to the more widely held notion that evolutionary changes in species are gradual and continuous over time.

Recombinant DNA
Genetically engineered segments of DNA that are inserted into a genome that would not naturally be found in that genome.

Reflective consciousness
As defined by John Hands in his book, *Cosmosapiens*, the property of an organism by which it is conscious of its own consciousness. That is, not only does it know but it also knows that it knows. Hands believes that *Homo sapiens* is the only species with this capability.

Reproductive isolation
The condition in which a species is unable (or less able) to interbreed with another species to produce viable offspring that, in turn, are able to have further offspring.

Ribosome
Structures within the cellular cytoplasm in which proteins are produced.

RNA
Ribonucleic acid – consists of four nucleotides: adenine, uracil, cytosine and guanine.

Sex chromosome
One of a pair of chromosomes that determines the sex of the individual. In humans, there is one pair of sex chromosomes (X for female, Y for male)

and 22 pairs of autosomal chromosomes. The X chromosome is relatively normal in size and gene count, whereas the Y chromosome is short, with relatively few coding genes.

Sexual selection
The mating preference of individuals of one sex of a species for certain physical or behavioral characteristics of the other sex.

Sila
A standard unit of volume of grain used as a form of value exchange in ancient times.

Singularitarian
An individual who promotes the notion that 'the singularity' can be achieved in a relatively short time (less than a century) and can be controlled for the good of humanity.

Singularity
As used in this book, the point at which human brain intelligence and computer artificial intelligence cannot be distinguished. Ray Kurzweil, Google's director of engineering, has popularized this notion.

Somatic cell
The type of cell that makes up all the tissues of an organism other than the germline cells. Changes in the DNA of somatic cells by mutation or genetic engineering are not passed on to progeny.

Speciation
The process of species origination in evolution.

Species
The basic unit of taxonomy in the classification of organisms. It typically represents the largest group of organisms that can produce viable offspring that, in turn, are capable of producing further offspring, but are reproductively isolated from all other species.

Species problem
The difficulty in defining what is a species, as reflected in the many different definitions.

Stem cell
A cell capable of developing into multiple cell types.

Synapse
A small gap between two neurons across which excitation takes place, with the participation of neurotransmitter chemicals.

Taxonomy

In biology, the study of the scientific classification of organisms into hierarchical groupings.

Vector

In genetic engineering, a biological vehicle used to deliver recombinant DNA into a cell. Viruses are commonly used vectors.

Zygote

A eukaryotic cell formed by the union of a male and female gamete (sperm and egg in humans). A zygote has the full complement of DNA to develop into a complete organism.

REFERENCES

1. Archaea are microscopic single-celled organisms that differ from bacteria in the nature of the cell walls and certain chemical processes. They are ubiquitous in nature, including within the human intestinal tract and on human skin, and are often found in extreme environments.
2. Gould, Steven. *Full House*, New York, Harmony Books, 1996, p.175.
3. Darwin, Erasmus. *Zoonomia; Or the Laws of Organic Life*, Vol. 1, Dublin: Project Gutenberg, 1794.
4. In truth, 'creationism' represents a broad spectrum of religious views about life, nature and evolution that could allow for evolution by natural selection as long as a divine creator started the whole process and somehow guides it.
5. There is a long-standing debate among paleontologists as to whether evolutionary changes are really gradual. The late renowned paleontologist Stephen Jay Gould popularized the term *punctuated equilibrium*, with which he argued that evolution proceeds in fits and starts. There are long periods of relative stability interspersed with relatively rapid and major changes leading to new species. This would explain the evolutionary gaps in the fossil record and suggest that changes occur in greater leaps. See https://phys.org/news/2020-06-popular-textbook-punctuated-evolution-debunked.html
6. Although Darwin and Wallace were unaware of Mendel's work, Mendel was certainly aware of Darwin's work and mentioned him in a couple of his publications.

7. The exception to this is red blood cells, which do not contain nuclei and therefore do not contain DNA or genes.

8. Conventionally, the term 'gene' usually refers to the genes that produce proteins, sometimes called 'coding genes.' Technically, the DNA sequences that are part of the epigenome are also genes, sometimes coding epigenomic proteins that only effect the expression of other genes and sometimes coding for RNA sequences that directly affect gene expression. Often these are referred to as non-coding genes. There are about 21,500 human coding genes and about 37,600 non-coding genes. https://www.nature.com/articles/d41586-020-02139-1

9. Mendel's publication showed the classic ratios of offspring phenotypes expected when breeding one parent with two dominant copies with another with two recessive copies of the same gene. The first generation consists of 100% dominant type phenotypes and the second generation has a ratio of 3/1 dominant/recessive phenotypes. Although those are correct, subsequent scholars looked at his data and concluded his data were too good to be true. That is, they didn't have the expected range of results that one would expect from normal statistical variation found in experiments and suspected him of fudging the results to prove his theories.

10. This is not entirely true. There are some genes that exhibit what is called 'co-dominance,' where both forms of the gene are expressed. For example, people with both the blood type gene for Type A blood and Type B blood will have a phenotype blood Type AB. There is also a situation called 'incomplete dominance' where, for example, the gene for a white flower and the gene for a red flower combined produce a pink flower.

11. For a nice explanation of this, see Livio, Mario. *Brilliant Blunders: From Darwin to Einstein.* New York: Simon & Schuster, 2013.

12. Each of the historical five mass extinctions eliminated 70–96% of species existing at those times, over periods ranging from a few thousand years to millions of years. The most recent was the asteroid impact 66 million years ago that wiped out the dinosaurs and most other species. In between the mass extinctions, species continue to go extinct at slower rates. Elizabeth Kolbert, in her book *The Sixth Extinction*, argues that we *Homo sapiens* are in the process of causing the sixth mass extinction.

13. Note that throughout this book, the standard punctuation convention will be used for species names in which the genus is capitalized and the species is not (except for its use in a title).

14. This view is not universally shared among anthropologists and taxonomists, some of whom argue that species are objective realities if one knows how to observe them. See Appendix 2 of my book, *What Comes After Homo Sapiens?*, for a more complete discussion of this issue.

15. De Queiroz, Kevin. "Species Concepts and Species Delimitation." *Systematic Biology* 56, issue 6 (2007): 879–886.

16. There is an international body that governs the *naming* of new animal species but does not get involved in determining whether or not a new finding constitutes a new species. It deals with all names throughout the entire animal kingdom. It is called the International Commission on Zoological Nomenclature, https://www.iczn.org. In contrast, there is the International Committee on Taxonomy of Viruses (https://talk.ictvonline.org) that is involved in both the naming and species designation of viruses. Viruses are not animals and are not considered alive.

17. Ibid, De Queiroz.

18. There are many published attempts to define life. The list is long and varied. Here is a sampling. Daniel Koshland's 'Seven Pillars of Life': program, improvisation, compartmentalization, energy, regeneration, adaptability, seclusion. (*Science* 295, no. 5563 (2002): 2215). Max Tegmark would define life as "a process that can retain its complexity and replicate." This definition could include future artificial intelligent computers, since they replicate information. In Tegmark's view, DNA replication is really information replication. (*Life 3.0*, Alfred A. Knopf, 2017). There is an attempt to define life as the result of quantum mechanics and create life in a quantum computer (https://www.sciencealert.com/scientists-simulate-artificial-life-in-quantum-algorithm-for-first-time). Byron Rees would say that anything that is conscious is alive (*The Fourth Age: Smart Robots, Conscious Computers, and the Future of Humanity*, Simon & Schuster, 2018), although there is great debate as to what is consciousness. Further, plants are not considered to be conscious, but they are certainly alive. Karl Friston, a well-known neuroscientist at University College London defines life as acting to minimize free energy. Not too many people, including myself, understand what that means. Stuart A. Kauffman states (*A World Beyond Physics – The Emergence and Evolution of Life*, Oxford University Press, 2019): "Life is a fundamentally new linking of non-equilibrium processes and boundary condition constraints on the release of energy into a few degrees of freedom that thus is thermodynamic work." (If you understand that sentence you are smarter than I.)

19. Prokaryotes reproduce by having their DNA divide into two identical copies allowing the organism to split into two new organisms. This process is called mitosis. Most eukaryotes reproduce sexually. In males to produce the sperm and females to produce the egg, a process called meiosis occurs. In meiosis, only half of the parental genetic material is incorporated into a sperm or egg. The full complement of DNA is recreated when a sperm fertilizes an egg. Sexually reproducing organisms also use mitosis to replicate all types of cells during growth and repair of damaged tissues.

20. https://www.zmescience.com/science/news-science/humans-live-longer-chimps-043242/

21. Other intermediate species from roughly the same periods whose fossils have been found in Africa are *Ardipithecus kadabba*, *Orrorin tugensis*, and *Sahelanthropis tchadensis*.

22. Pavid, Katie. "*Australopithecus afarensis*: Human Ancestors Had Slow-growing Brains Just Like Us." *Natural History Museum, Science News*, April 1, 2020. https://www.nhm.ac.uk/discover/news/2020/april/australopithecus-afarensis-had-slow-growing-brains.html

23. For a longer discussion of this topic, see McHenry, Henry, "Origin of Human Bipedality," *Evolutionary Anthropology,* 13(2004): 116-119.

24. Tattersall, Ian and Desalle, Rob. *The Accidental Homo Sapiens: Genetics, Behavior, and Free Will.* New York: Pegasus Books, 2019, p. 95.

25. Thanks to the work of Svante Pååbo and his team at the Max Planck Institute for Evolutionary Anthropology in Leipzig Germany, we can now extract enough DNA from some fossils of extinct species such as Neanderthals to perform complete genomic analyses for comparison to living species. In fact, such analysis is the only way we know of the existence of Denisovans.

26. This set of genes is labeled NOTCH2NL.

27. Pollard, Katherine. "What Makes Us Human?" *Scientific American* (May 2009): 44–49.

28. This gene is labeled NYH16.

29. For most of the past century, it was thought that the first *Homo sapiens* arrived in the Americas over a Bering Strait land bridge and proceeded south after a channel opened up in the ice around 13,500 years ago (the so-called Clovis people). There is some recent archeological evidence from Idaho that the first North American humans might have come as early as 16,000 years ago, which would be before the melting of the ice cover that prevented the northern land migration. This implies a

migration over sea by boat from Asia and down along the Pacific Coast. Further, the stone tool artifacts are similar to those found in Japan, which may suggest that the first North American humans originated in Japan. L. Davis, D Madsen, L Becerra-Valdivia, et. al, "Late Upper Paleolithic Occupation at Cooper's Ferry, Idaho, USA ~16,000 Years Ago," Science, August 2019, doi: 10.1126/science.aax9830

30. Ghost genetic tracings occur when examination of the genomes of extinct or ancient humans reveal patterns of mutations that do not belong to any known human species. They are referred to as ghost species, which are hypothesized or speculated other human species for which we have no fossil evidence or earlier human species whose genomes are not known.

31. Sometimes all of the various *Homo* species from that era that may have been the immediate precursors to *Homo sapiens* are referred to as 'archaic' humans.

32. In the past 1 million years, there have been at least nine human species other than *Homo sapiens* depending on various fossil interpretations: *Homo erectus, Homo heidelbergensis, Homo naledi, Homo helmie, Homo antecessor, Homo denisova, Homo neanderthalensis, Homo floresiensis, and Homo luzonensis*. Some of these are grouped together simply as 'archaic' Homo sapiens. Since we don't have DNA from most of these fossils, we don't know which ones, if any, correspond to so-called 'ghost' ancestors found in various genomic analyses.

33. McBrearty, Sally and Brooks, Allison. "The Revolution that Wasn't: A New Interpretation of the Origin of Modern Human Behavior." *Journal of Human Evolution* 39, Issue 5 (2000): 453–563.

34. Cepelewicz, Jordana. "Fossil DNA Reveals New Twists in Modern Human Origins." *Quanta Magazine*, August 29 2019. https://www.quantamagazine.org/fossil-dna-reveals-new-twists-in-modern-human-origins-20190829/.

35. Hubisz, M., Williams, A., Siepel, A. "Mapping Gene Flow Between Ancient Hominins Through Demography-Aware Inference of the Ancestral Recombination Graph." *PLOS Genetics*, 16(8) (2020), e1008895. https://journals.plos.org/plosgenetics/article?id=10.1371/journal.pgen.1008895

36. Gibbons, Ann. "Strange Bedfellows for Human Ancestors." *Science Magazine*, 367, issue 6480 (February 21, 2020): 838–839.

37. We can date the onset of *Homo sapiens* clothing by the study of the genomes of our body lice which only live in our clothing. Dating techniques have determined these lice emerged between 80,000 and

170,000 years ago – before our ancestors left Africa to replace all other humans. They left clothed. There is also good research evidence that *Homo sapiens* at that time used fitted clothing whereas the Neanderthals did not: *Journal of Anthropological Archaeology*, 44 (part B): 235–246, December 2016.

38. Degioanni, Anna, Bonenfant, Christophe, Cabut, Sandrine, et. al. "Living on the edge: Was demographic weakness the cause of Neanderthal demise?" *PLOS One*, May 29, 2019. https://doi.org/10.1371/journal.pone.0216742

39. Timmerman, Axel, "Quantifying the Potential Causes of Neanderthal Extinction: Abrupt Climate Change Versus Competition and Interbreeding." *Quaternary Science Reviews*, 238 (2020): 106331. http://dx.doi.org/10.1016/j.quascirev.2020.106331

40. Tattersall, Ian. Ibid, p.124.

41. David Reich, *Who We Are and How We Got Here: Ancient DNA and the New Science of the Human Past*. New York: Pantheon Books, 2018.

42. Allen, Paul and Greaves, Mark. "The Singularity Isn't Near." *MIT Technology Review*, October 12, 2011. https://www.technologyreview.com/s/425733/paul-allen-the-singularity-isnt-near/

43. There is one isolated fossil find about 3.3 million years ago in Kenya that also contains what appears to be a deliberately created stone flake which could be considered the first tool. It is called Lomekwi (named after the site in which it was found) and probably made by an *Australopithecus* species.

44. Mayer, D., Groman-Yaroslavski, I., Bar Yosef, O. "On Holes and Strings: Earliest Displays of Human Adornment in the Middle Palaeolithic." *PLOS One*, 2020. https://doi.org/10.1371/journal.pone.0234924

45. I said that genes can only be passed from parent to child "in humans." That is also true in all plants and other animals. That is not true in prokaryotes. In bacteria and archaea, the main form of gene transfer is from parent to child through mitosis reproduction. However, prokaryotes have another form of horizontal or lateral gene transfer where DNA is passed from one organism to another through a connection called conjugation. The genes for antibiotic resistance are an example of such lateral gene transfer not only within a bacterial species, but between bacterial species. In fact, there are many other types of lateral or horizontal gene transfer throughout the Eukarya domain as well.

46. In fact, Paul Romer of NYU received the 2018 Nobel Memorial Prize in Economic Sciences for proving this concept. https://voxeu.org/article/new-ideas-about-new-ideas-paul-romer-nobel-laureate

47. Reich, David. Ibid loc. 2231 of e-book.

48. Harari, Yuval. *Sapiens: A Brief History of Mankind.* New York: Harper, 2014.

49. A better tool might have been an fMRI (functional magnetic resonance image), which shows which neural networks in the brain are active during specific mental activities.

50. Tattersall, Ian. Ibid, p.124.

51. https://quotes.thefamouspeople.com/alan-turing-4223.php

52. These four computer scientists were John McCarthy from Dartmouth College, Marvin Minsky from Harvard College, Nathaniel Rochester from IBM, and Claude Shannon from Bell Telephone Laboratories in their "Proposal for the Dartmouth Summer Research Project on Artificial Intelligence," 1955. http://raysolomonoff.com/dartmouth/boxa/dart564props.pdf

53. For this discussion, "human" intelligence will be assumed to be "*Homo sapiens*" intelligence. Although other species of humans were intelligent also, as discussed in *Chapter 2,* we may be the most intelligent, no matter how that is defined. In any case, we don't have access to any other human species' brains to study.

54. Charles Babbage is credited with designing the first digital computer in the 19th century. His device was often referred to as a 'thinking machine.'

55. For a review of the science and theories of consciousness, see Scientific American e-books edition, *The Science of Consciousness*, September 2019.

56. Tomasello, Michael and Carpenter, Malinda. "Shared Intentionality." *Developmental Science* 10, no.1 (2007): 121–125.

57. Greene, Brian. *Until the End of Time: Mind, Matter, and Our Search for Meaning in an Evolving Universe.* New York: Alfred A. Knopf, 2020.

58. There is an artificial-intelligence-based robot named AI-DA (named after the first computer programmer Ada Lovelace) that creates Picasso-like abstract art. See https://www.metro.news/connect-meet-ai-da-the-worlds-first-robot-artist/1614828/

59. The website, Quote Investigator (https://quoteinvestigator.com/2011/11/05/computers-useless/), researched the true origin of this quote and concluded that it could reasonably be attributed to Picasso.

60. This also raises the question as to what is really being measured by so-called intelligence tests.

61. Actually, these chess programs are not totally brute force. They typically pre-program the early moves called openings according

to a portfolio of long-established *openings*. There are also other programming shortcuts, which attempt to reduce the number of potential possibilities that it needs to explore. However, the main growth in performance of chess software was related to the growth in computing speed.

62. Lasker states: "...simply enumerating all possible variations ... I contend, is quite useless. It certainly, as the experience of many centuries indisputably shows, would by no means exclude the possibility of committing grave errors." And later, "commons sense triumphs over trickery."

63. Ironically, the mirror image of IA seems to work with AI systems. For example, in 2009 – 12 years after his loss to Deep Blue – Gary Kasparov learned to his chagrin that the IBM team had built into the Deep Blue software some arbitrary delays in making its moves simply to try to confuse Kasparov during their famous match in 1997. These random delays caused Kasparov to ponder a little longer as to what might be behind the computer's move. This has nothing to do with AI but rather simply adding some 'human' to the machine software was better than the machine software alone.

64. Kasparov, Gary. *Deep Thinking: Where Machine Intelligence Ends and Human Creativity Begins*. New York: Public Affairs, 2017.

65. Moravcik, Matej, Schmid, Martin, Burch, Neil, et. al. "DeepStack: Expert-level Artificial Intelligence in Heads-up No-limit Poker." *Science* 356, issue 6337 (May 5, 2007): 508-513. DOI: 10.1126/science.aam6960

66. Brown, Noam and Sandholm, Tuomas. "Superhuman AI for Multiplayer Poker." *Science* 365, issue 6456 (July 11, 2019): 885-890. DOI:10.1126/science.aay2400

67. For an explanation of 'donk' bets, see https://www.pokerstarsschool.com/strategies/donk-betting-what-it-is-and-when-to-do-it/1047/

68. Kamlish, Isaac, Chocron, Isaac and McCarthy, Nicholas. "SentiMATE: Learning to Play Chess Through Natural Language Processing." https://arxiv.org/pdf/1907.08321.pdf.

69. https://arxiv.org/pdf/1912.06680.pdf

70. A recent definition of AI as reported in *Science Magazine* is "A growing resource of interactive, autonomous, self-learning agency, which enables computational artifacts to perform tasks that otherwise would require human intelligence to be executed successfully." (Taddeo, M. and Floridi, L. "How AI Can be a Force for Good." *Science* 361, issue 6404: 751–752, August 24, 2018). I find that definition so convoluted as to be impractical as a useful reference point.

71. As an example, see Gee, Sue. "Lovelace 2.0 Test – An Alternative Turing Test," November 24, 2014. https://www.i-programmer.info/news/105-artificial-intelligence/7999-lovelace-20-test-an-alternative-turing-test.html

72. Saygin, Ayse, Cicekli, Ilyas, and Akman, Varil. "Turing Test: 50 Years later." *Minds and Machines* 10, issue 4 (2000): 463–518.

73. Laguarta, Jordi, Huerto, Ferran, Subirana, Brian. "COVID-19 Artificial Intelligence Diagnosis Using Only Cough Recordings." *IEEE Open Journal of Engineering in Medicine and Biology*, September 2020, doi: 10.1109/OJEMB.2020.3026928.

74. When I wrote *What Comes After Homo Sapiens?* it was thought that *C. elegans* had fewer than 2,000 genes. Later reports have revised that up to ten times that number. No, the worm is not evolving that quickly – just our sequencing techniques.

75. Callaway, Ewen. "'It Will Change Everything': DeepMind's AI Makes Gigantic Leap in Solving Protein Structures." *Nature*, https://www.nature.com/articles/d41586-020-03348-4. https://doi.org/10.1038/d41586-020-03348-4

76. Vinyals, Oriol, Babuschkin, Igor,Czarnecki, Wojciech, et. al. "Grandmaster Level in StarCraft II Using Multi-agent Reinforcement Learning," *Nature* 575(2019): 350–354. doi:10.1038/s41586-019-1724-z

77. Brown, T., Mann B., Ryder, N., et. al. "Language Models are Few-Shot Learners." https://arxiv.org/abs/2005.14165

78. Du Sautoy, Marcus. *The Creativity Code: Art and Innovation in the Age of AI*, Cambridge: The Belknap Press of Harvard University Press, 2019.

79. Wee, Sui-Lee. "China is Collecting DNA from Tens of Millions of Men and Boys, Using U.S. Equipment." *New York Times*, June 17, 2020. https://www.nytimes.com/2020/06/17/world/asia/China-DNA-surveillance.html?campaign_id=57&emc=edit_ne_20200617&instance_id=19481&nl=evening-briefing®i_id=56213537&segment_id=31174&te=1&user_id=d5565cd06ef97d73c96909b2c44c1e3b

80. Larson, Jeff, Mattu, Surya, Kirchner, Lauren, et. al. "How We Analyzed the COMPAS Recidivism Algorithm." *Pro Publica*, May 23, 2016. https://www.propublica.org/article/how-we-analyzed-the-compas-recidivism-algorithm

81. Obermeyer, Ziad, Powers, Brian, Vogeli, Christine, et.al., "Dissecting Racial Bias in an Algorithm Used to Manage the Health of Populations." *Science Magazine* 366, issue 6464 (October 25, 2019): 447–453.

82. Zech, John, Badgeley, Marcus, Liu, Manway, et.al., "Variable generalization performance of a deep learning model to detect pneumonia in chest radiographs: A cross-sectional study." *PLOS medicine*, November 6, 2018. https://doi.org/10.1371/journal.pmed.1002683

83. Lazar, David, Kennedy, Ryan, King, Gary, et. al., "The Parable of Google Flu: Traps in Big Data Analysis." *Science*, issue 6176 (March 14, 2014): 1203–1205.

84. O'Brian, M. "Retractions and Controversies Over Coronavirus Research Show That the Process of Science is Working as it Should." https://theconversation.com/retractions-and-controversies-over-coronavirus-research-show-that-the-process-of-science-is-working-as-it-should-140326

85. Röösli, E., Rice, B., Hernandez-Boussard, T. "Bias at Warp Speed: How AI Might Contribute to the Disparities Gap in the Time of COVID-19." *JAMIA*, ocaa210 (2020). https://doi.org/10.1093/jamia/ocaa210

86. https://www.aei.org/economics/what-atms-bank-tellers-rise-robots-and-jobs/

87. Muro, Mark, Whiton, Jacob and Maxim, Robert. "What Jobs are Affected by AI." *Metropolitan Policy Program at Brookings*, November 2019. https://www.brookings.edu/wp-content/uploads/2019/11/2019.11.20_BrookingsMetro_What-jobs-are-affected-by-AI_Report_Muro-Whiton-Maxim.pdf

88. Berriman, Richard, Hawksworth, John. "Will Robots Steal our Jobs?" *Price-Waterhouse UK Economic Outlook*, March 2017. https://www.pwc.co.uk/economic-services/ukeo/pwcukeo-section-4-automation-march-2017-v2.pdf

89. Grace, Katja, Salvatier, John, Dafoe, Allan, et. al. "When Will AI Exceed Human Performance? Evidence from AI Experts." May 2018. https://arxiv.org/abs/1705.08807

90. For a fascinating read of a fictional (hopefully) view of a possible dystopian future of AI based on our current non-fictional understanding of AI see the book *Burn-In* by P.W. Singer and August Cole.

91. The IEEE (Institute of Electrical and Electronic Engineers) is the largest technical professional organization in the world. The IEEE Global Initiative on Ethics of Autonomous and Intelligent Systems has produced a report addressing these issues: *Ethically Aligned Design: A Vision for Prioritizing Human Well-being with Autonomous and Intelligent Systems*, Version 2. IEEE, 2017. http://standards.ieee.org/develop/indconn/ec/autonomous_systems. html

92. Drum, Kevin. "You Will Lose Your Job to a Robot – And Sooner Than You Think." *Mother Jones*, November/December 2017 Issue. https://www.motherjones.com/politics/2017/10/you-will-lose-your-job-to-a-robot-and-sooner-than-you-think/

93. https://www.youtube.com/watch?v=xhpXU0x5894

94. Google Brain is Google's in-house AI think tank and research team. https://ai.google/research/teams/brain

95. Real, Esteban, Liang, Chen, So, David. "AutoML-Zero: Evolving Machine Learning Algorithms from Scratch." *arXiv.2003.03384*. https://arxiv.org/abs/2003.03384

96. Hudson, Matthew. "Hackers Easily Fool Artificial Intelligences." *Science* 361, issue 6399 (July 2018): 215.

97. https://www.janes.com/defence-news/news-detail/heron-systems-ai-defeats-human-pilot-in-us-darpa-alphadogfight-trials

98. Turek, Matt. "Explainable Artificial Intelligence (XAI)." *DARPA*. https://www.darpa.mil/program/explainable-artificial-intelligence

99. Accenture Labs. "Understanding Machines: Explainable AI." https://www.accenture.com/_acnmedia/pdf-85/accenture-understanding-machines-explainable-ai.pdf#zoom=50

100. Bostrom, Nick. *Superintelligence: Paths, Dangers, Strategies*. Oxford: Oxford University Press, 2014.

101. Some of these other organizations include The Machine Intelligence Research Institute (https://intelligence.org), The Foresight Institute (www.foresight.org), The Future of Life Institute (https://futureoflife.org/team/), Humanity+ (www.humanityplus.org), AI4People (https://www.eismd.eu/ai4people/), Leverhulme Centre for the Future of Intelligence (http://lcfi.ac.uk), K&L Gates Endowment for Ethics and Computational Technologies (https://www.cmu.edu/news/stories/archives/2016/november/gift.html), Ethics and Governance of Artificial Intelligence Fund (https://www.knightfoundation.org/press/releases/knight-foundation-omidyar-network-and-linkedin-founder-reid-hoffman-create-27-million-fund-to-research-artificial-intelligence-for-the-public-interest), The Future of Humanity Institute (https://www.fhi.ox.ac.uk), Partnership on AI (https://www.partnershiponai.org), OpenAI (https://openai.com) and The Centre for the Study of Existential Risk (https://www.cser.ac.uk).

102. Good, Irving. "Speculations Concerning the First Ultraintelligent Machine," in Franz L. Alt and Morris Rubinoff, *Advances in Computers*, Vol 6, New York: Academoic Press, Inc., 1965.

103. For example, see his *Marooned in Realtime*.

104. Presented at the VISION-21 Symposium sponsored by NASA Lewis Research Center and the Ohio Aerospace Institute, March 30-31, 1993. https://archive.org/stream/NASA_NTRS_Archive_19940022855/NASA_NTRS_Archive_19940022855_djvu.txt

105. Heylighen, Francis. "Return to Eden? Promises and Perils on the Road to a Global Superintelligence," in Ben Goertzel and Ted Goertzel (eds.), *The End of the Beginning: Life, Society and Economy on the Brink of the Singularity*. Los Angeles: Humanity+ Press, 2015.

106. Heylighen, Francis. "The Global Superorganism: An Evolutionary-cybernetic Model of the Emerging Network Society." *Journal of Social Evolution and History*, 6, no.1 (2007): 58–119. http://pespmc1.vub.ac.be/Papers/Superorganism.pdf

107. Kurzweil, Ray. *The Singularity is Near*. New York: Viking, 2005.

108. Reimann, Michael, Gevaert, Michael, Shi, Ying, et. al. "A Null Model of the Mouse Whole-neocortex Micro-connectome." *Nature Communications* 10: Article number 3903 (August 29, 2019). https://rdcu.be/b15Co

109. BRAIN 2025 Report, National Institutes of Health, The BRAIN Initiative, June 5, 2014. https://braininitiative.nih.gov/strategic-planning/brain-2025-report

110. https://braininitiative.nih.gov/strategic-planning/acd-working-groups/brain-initiative-20-cells-circuits-toward-cures

111. Glasser, Matthew, Smith, Stephen, Marcus, Daniel, et. al. "The Human Connectome Project's Neuroimaging Approach." *Nature Neuroscience* 19 (August 26, 2016): 1175–1187. https://www.nature.com/articles/nn.4361

112. Jiang, Linxing, Stocco, Andrea, Losey, Darby, et. al. "BrainNet: A Multi-person Brain-to-Brain Interface for Direct Collaboration Between Brains." *Sci Rep* 9, article no. 6115 (April 16, 2019) doi:10.1038/s41598-019-41895-7

113. Durham, Emily, "Team Receives $19M from DARPA to Create Noninvasive Brain Interface." https://www.cmu.edu/news/stories/archives/2019/may/darpa-brain-interface.html

114. https://www.darpa.mil/news-events/2019-05-20

115. O'Doherty, Joseph, Lebedev, Mikhail, Ifft, Peter, et. al. "Active Tactile Exploration Using a Brain-Machine-Brain Interface." *Nature* 479 (2011): 228–231.

116. Raspopovic, Stanisa. "Advancing Limb Neural Prostheses." *Science* 370, Issue 6514, (October 16, 2020): 290–291

117. Pandarinath, Cheatham and Ali, Yahia. "Brain Implants that Let You Speak Your Mind." *Nature* 568 (April 29, 2019): 466–467.

118. Liu, Jia, Fu, Tian-Ming, Cheng, Zengguang, et. al. "Syringe Injectable Electronics." *Nature Nanotechnology* 10 (2015): 629–636.

119. See https://www.fda.gov/regulatory-information/search-fda-guidance-documents/breakthrough-devices-program

120. Implanted device measures 0.9 inches in diameter and 0.3 inches in depth.

121. This article was written by a team of neuroscientists headed by Robert A. Freitas, Jr., at the Institute for Molecular Manufacturing in Palo Alto, CA. See: Martins, Nuno, Angelica, Amara, Chakravarthy, Krishnan, et. al. "Human/Brain Cloud Interface." *Frontiers in Neuroscience*, 29 March 2019. https://www.frontiersin.org/articles/10.3389/fnins.2019.00112/full#:~:text=%20Human%20Brain%2FCloud%20Interface%20%201%20Introduction.%20There,to%20and%20from%20living%20human%20brains...%20More%20

122. For those scientifically inclined, there is a good review article of this type of research: B Esteban-Fernandez de Avila, W Gao, E Karshalev, et. al. "Cell-like Micromotors." *Accounts of Chemical Research*, May 2018, DOI:10.1021/acs.accounts.8b00202

123. Felfoul, Ouajdi, Mohammadi, Mahmood, Taherkhani, Samira, et. al. "Magneto-aerotactic bacteria deliver drug-containing nanoliposomes to tumour hypoxic regions." *Nature Nanotechnology* 11 (2016): 941–947. doi:10.1038/nnano.2016.137

124. Kriegman, Sam, Blackiston, Douglas, Levin, Michael, et. al. "A Scalable Pipeline for Designing Reconfigurable Organisms." *PNAS* 117 issue 4 (2020): 1853–1859. https://doi.org/10.1073/pnas.1910837117

125. Allen, Paul and Greaves, Mark, Ibid.

126. Pein, Corey. "The Singularity is Not Near: The Intellectual Fraud of the 'Singularitarians.' " *Salon*, May 13, 2018. https://www.salon.com/2018/05/13/the-singularity-is-not-near-the-intellectual-fraud-of-the-singularitarians/

127. There is one experiment in mice where researchers were able to create what appears to be a false memory by stimulating certain brain cells. Specifically, they were able to make the mouse fear a situation as though it had been previously conditioned to fear it – even though no such prior conditioning had occurred. This was a very complex experiment that I haven't seen replicated, nor anything like it, regarding knowledge or capabilities. See: S Ramirez, S, X Liu, P Lin, et. al. "Creating a False Memory in the Hippocampus." *Science* 341(6144): 387, 2013

128. Boudry, Maarten. "Human Intelligence: Have We Reached the Limit of Knowledge?" *The Conversation*, October 11, 2019. https://www.theconversation.com/human-intelligence-have-we-reached-the-limit-of-knowledge-124819

129. Joy, Bill. "Why the Future Doesn't Need Us." *Wired Magazine*, April 2000, https://www.wired.com/2000/04/joy-2/

130. This sounds simple, but in reality, was quite complicated. Human insulin is a protein consisting of 51 amino acids divided into two chains of 21 and 30 amino acids which are chemically linked together. Knowing the amino acid sequence of each chain, the researchers at Genentech could reverse engineer the exact DNA sequence needed to produce each chain. Using DNA splicing techniques, they pieced together small snippets of DNA to create the full sequence of each chain. They then inserted each newly created "gene" into the E. coli DNA using recombinant DNA techniques, one strain of bacteria for each of the two chains. After letting the bacteria replicate and produce huge quantities of each insulin chain, they extracted the proteins, linked them together chemically and, *voila*, human insulin was produced!

131. A 'trait' of any organism that is the manifestation of a gene or set or genes is called the *phenotype*. The actual gene sequence is called the *genotype*. Phenotypes may be normal traits or characteristics of an organisms (such as eye color) or abnormal characteristics as in the many genetic diseases. Finding the exact gene (genotype) associated with any phenotype can be done in several ways. One is to compare large numbers of individual genomes between individuals with and without a given phenotype to determine where in the genome there is a consistent pattern of DNA associated with each. Another method is if there is some specific known protein associated with a particular phenotype, for example, sickle hemoglobin, the amino acid sequence of that protein can be determined. This, in turn, would indicate the DNA sequence that would be required to create that protein. That sequence, or gene, can then be searched for in the genome.

132. This genetic engineering experiment in flies was done only in a laboratory and in no way created a competitor to the threatened monarch butterfly which requires milkweed for survival. The purpose was to better understand how the evolution of resistance to toxic chemicals occurs in nature. Chemicals in milkweed are toxic to most predators of the monarch butterflies which provides a natural form of protection for them. https://phys.org/news/2019-10-scientists-recreate-flies-mutations-monarch.html

133. McKie, Robin. "Block on GM rice 'has cost millions of lives and led to child blindness.'" *The Guardian*, October 26, 2019. https://www.theguardian.com/environment/2019/oct/26/gm-golden-rice-delay-cost-millions-of-lives-child-blindness

134. Liu, Wusheng, Rudis, Mary, Cheplick, Matthew, et. al. "Lipofection-mediated Genome Editing Using DNA-free Delivery of the Cas9/gRNA Radionucleoprotein into Plant Cells." *Plant Cell Reports* 39 (2020): 245–257

135. Culver, K., Anderson, W., Blaese, R. "Lymphocyte Gene Therapy." *Human Gene Therapy* 2, No. 2 (1991): 10.

136. Dave, U., Jenkins, N., Copeland, N. "Gene Therapy Insertional Mutagenesis Insights," *Science* 303 (2004): 333.

137. https://www.who.int/genomics/public/geneticdiseases/en/index2.html

138. For a review of available commercial genetic engineering-based drugs, see https://www.scientificamerican.com/article/gene-therapy-arrives/

139. One start-up company, Cell Vault, is providing the ability for any healthy person to have their blood drawn and the T-cells extracted and frozen for potential use in case they need it in the future.

140. https://www.fda.gov/news-events/press-announcements/fda-approves-first-cell-based-gene-therapy-adult-patients-relapsed-or-refractory-mcl

141. For example, a gene called ROBO1 has been identified that increases the volume of an area of the human brain known to be important in mathematical ability. Skeide, Michael, Wehrmann, Katharina, Emani, Zahra, et. al. "Neurobiological Origins of Individual Differences in Mathematical Ability." *PLOS Biology* 18 (10): e3000871, October 22, 2020. https://doi.org/10.1371/journal.pbio.3000871

142. There are two general types of viruses: DNA-based and RNA-based. When the bacterium needs a defense against an RNA-based virus, its CRISPR component contains the DNA sequence that corresponds to the RNA of the virus genome.

143. Jenik, Martin, Chylinski, Krzysztof, Fonfara, Ines, et. al. "A Programmable Dual-RNA-guided DNA Endonuclease in Adaptive Bacterial Immunity." *Science* 337, issue 6096 (2012): 816–821.

144. Actually, another researcher, Virginijus Siksnys of Vilnius University in Lithuania, discovered the same potential for CRISPR at about the same time, but his paper was initially rejected by the journal to which he submitted his findings. It was later published in another journal. Many feel that he deserves the same credit as Jennifer Doudna and Emmanuelle Charpentier for the discovery of the genetic engineering potential of CRISPR.

145. Pennisi, Elizabeth. "The CRISPR Craze." *Science* 341, issue 6148 (2013): 833–836.

146. This discussion of repairing the double-stranded break in DNA is a good example of the tremendous complexity of the genome and genetic engineering. There are two general ways that the DNA break may be repaired by CRISPR called non-homologous end-joining (NHEJ) and homology directed repair (HDR). NHEJ is generally how cells naturally fix breaks and is used by CRISPR to 'knock out' a harmful gene. HDR is generally how CRISPR is used to add corrective DNA during gene therapy or 'knock-in' a good gene. However, to truly understand and describe these repair mechanisms is beyond the scope of this book. It's one of the reasons we will need AI to make this all work.

147. Mullin, E. "Scientists Edited Human Embryos in the Lab, and It Was a Disaster." https://onezero.medium.com/scientists-edited-human-embryos-in-the-lab-and-it-was-a-disaster-9473918d769d

148. Fararrelli, Leslie. "CRISPR, Cancer and p53." *Science Signaling*, Vol 11, Issue 539, July 17, 2018. DOI: 10.1126/scisignal.aau7344

149. Statement for the Record, Worldwide Threat Assessment of the US Intelligence Community, Senate Armed Services Committee, James R. Clapper, Director of National Intelligence, February 9, 2016, https://www.dni.gov/files/documents/SASC_Unclassified_2016_ATA_SFR_FINAL.pdf

150. Ribeil, Jean-Antoine, Haycein-Bay-Abina, Salima, Payen, Emmanuel, et. al. "Gene Therapy in a Patient with Sickle Cell Disease." *NEJM* 376 (2017): 848–855.

151. Nuffield Council on Bioethics. "Genome Editing: An Ethical Review." September 30, 2016. https://nuffieldbioethics.org/publications/genome-editing-an-ethical-review

152. Shaw, David. "The Consent Form in the Chinese CRISPR Study: In Search of Ethical Gene Editing." *Journal of Bioethical Inquiry* (2020). https://doi.org/10.1007/s11673-019-09953-x

153. Musunuru, Kiran. "Opinion: We Need to Know What Happened to CRISPR Twins Lulu and Nana." *MIT Technology Review*, December 3, 2019. https://www.technologyreview.com/s/614762/crispr-baby-twins-lulu-and-nana-what-happened/

154. https://phys.org/news/2020-09-biologists-genetic-neutralize-gene.html

155. This possibility is discussed in George Church and Ed Regis' book, *Regenesis*. If you think that is bizarre or scary, consider this possibility.

Stem cells are very early cells in our cellular development cycle that differentiate into other types of cells as we develop in the embryo. For example, stem cells in the bone marrow develop into white blood stem cells which eventually branch into different types of white blood cells. A fertilized egg is the ultimate stem cell, which develops into all other cells. We now know how to reverse that process. That is, we can take a mature adult cell and coax it back into becoming a stem cell from virtually any cell in your body. Further, we have learned how to then control those stem cells into developing into virtually any other cell in your body.

Theoretically then, we can take some of the cells from your skin (or someplace else) and, in a petri dish, convert them into stem cells. We could then take some of these stem cells and coax them into becoming eggs and others into becoming sperm. We could then have one of these sperm and eggs unite *in vitro* into a fertilized egg and implant that into any female. The resulting newborn would be created entirely from your DNA. But it would not be your clone. It would be more like you than your normal children, but still not a clone. It would simply be a different mixture of the two copies you have already of each gene.

Using a similar technique, two women could have a male or female child together that would be a mixture of their genetic material the same as any other couple. Does all of this remind you of George Orwell's famous book *1984*? Actually, this is more advanced. And it is all just speculation. It has never been done. I doubt it will ever be done in humans. But who knows?

156. Cohen, Jon. "Neanderthal Brain Organoids Come to Life." *Science* 360, issue 6395 (2018): 1284.

157. Burrell, Teal. "Scientists Put a Human Intelligence Gene into a Monkey." *Discover Magazine*, December 29, 2019. https://www.discovermagazine.com/mind/scientists-put-a-human-intelligence-gene-into-a-monkey-other-scientists-are

158. Heide, Michael, Haffner, Christiane, Murayama, Ayako, et. al. "Human Specific ARHGAP11B Increases Size and Folding of Primate Neocortex in the Fetal Marmoset." *Science* 369, issue 6503 (2020): 546–550, DOI: 10.1126/science.abb2401

159. Gibson, Daniel, Benders, Gwynedd, Andrews-Pfannkoch, Cynthiak et.al. "Complete Chemical Synthesis, Assembly, and Cloning of a Mycoplasma Genitalium Genome." *Science* 319, issue 5867 (2008): 1215–1220. Doi:10.1126/science.1151721

160. Gibson, Daniel, Glass, John, Lartigue, Carole, et. al. "Creation of a Bacterial Cell Controlled by a Chemically Synthesized Genome." *Science* 329, issue 5987 (2010): 52–56.

161. There are four nucleotides in DNA. Any three of them constitutes a codon for an amino acid. There can be 64 combinations of these triplets. Since there are only 20 amino acids plus some "punctuation marks" designating the start and stop of a gene coding for proteins, there are extra triplets allowing most amino acids to have more than one codon. These same 64 triplets can be used to code for numbers and letters of the alphabet when used in areas of the genome that are not between start and stop sequences indicating a coding gene.

162. Kriegman, Sam, Blackiston, Douglas, Levin, Michael, et. al. "A Scalable Pipeline for Designing Reconfigurable Organisms." *Proceedings of the National Academy of Sciences* (2020). https://doi.org/10.1073/pnas.1910837117

163. Zhang, Yorke, Lamb, Brian, Feldman, Aaron, et. al. "A Semi-synthetic Organism Engineered for the Stable Expansion of the Genetic Alphabet." *Proceedings of the National Academy of Sciences* (January 2017). https://doi.org/10.1073/pnas.1616443114

164. Hoshika, Shuichi, Leal, Nicole, Kim, Myong-Jung, et. al. "Hachimoji DNA and RNA: A Genetic System with Eight Building Blocks." *Science* 363, issue 6429 (2019): 884–887.

165. https://www.idlehearts.com/2417083/knowledge-is-telling-the-past-wisdom-is-predicting-the-future

166. In classical computing, a digital computer 'bit' can be only in one of two states: 0 or 1. A qubit, which is the quantum computer's equivalent to a bit, takes advantage of the property of quantum mechanics called superimposition. Superimposition states that the qubit can exist in both states, 0 and 1, simultaneously. For those who understand all of this (and I do not include myself among them), this property of qubits allows a quantum computer to be vastly more powerful than a classical digital computer. How much more powerful? In October 2019, Google announced that its quantum computer took just 200 seconds to do a calculation that the most powerful digital computer would have taken 10,000 years to do. Currently, quantum computers can only operate at very low temperatures and are quite difficult to operate. This will change in this century.

167. An interesting GWAS study was done during the COVID pandemic that demonstrated that people with Type A blood type were more likely to respond to the infection with a severe illness and people with Type O blood type with less sever illness. https://www.nejm.org/doi/full/10.1056/NEJMoa2020283?query=TOC

168. Fehlmann, Tobias, Kahraman, Mustafa, Ludwig, Nicole, et. al. "Evaluating the Use of Circulating MicroRNA Profiles for Lung Cancer Detection in Symptomatic Patients." *JAMA Oncology*, March 5, 2020. doi:10.1001/jamaoncol.2020.0001

169. The investment community is betting on our ability to detect cancers of all types long before there are any clinically detectable manifestations in this manner. That is, in some cases, we will even detect the cancer earlier than stage 1. Just by drawing a blood sample. Two such companies, Grail and Freenome, have had huge investments by the end of 2019 and are in long-term clinical studies. There will be others.

170. Sarkar, Tapash, Quarta, Marco, Mukherjee, Shravani, et. al. "Transient Non-integrative Expression of Nuclear Reprogramming Factors Promotes Multifaceted Amelioration of Aging in Human Cells." *Nature Communications*, March 24, 2020. https://doi.org/10.1038/s41467-020-15174-3

171. In decreasing order of population, they are Shanghai, Beijing, Guangzhou, Tianjin and Shenzhen.

172. It is common for a coding gene to produce more than one protein and/or RNA molecules. This is done through a mechanism called *alternative splicing*.

173. https://med.nyu.edu/pathology/research/featured-research-archives/maternalfetal-immune-interactions-why-doesnt-mother-reject

174. Lavialle, Christian, Cornelis, Guillaume, Dupressoir, Anne, et. al. "Paleovirology of 'syncytins', Retroviral *env* Genes Exapted for a Role in Placentation." *Philos Trans R Soc Lond B Biol Sci* 368, 1626 (2013): 20120507. doi.org/10.1098/rstb.2012.0507

175. Foley, Robert. *Humans Before Humanity*. Hoboken: Wiley-Blackwell, 1995.

176. There are early studies in mice demonstrating possible genetic engineering approaches to obesity.

177. Varki, A, Gagneux, P, Multifarious Roles of Sialic Acids in Immunity, *Ann N Y Acad Sci* 1253(1) (2012): 16–36, https://www.ncbi.nlm.nih.gov/pmc/articles/PMC3357316/

INDEX

Note: Page numbers followed by *f* and *t* indicate figures and tables, respectively.

B

H

N

O

P

CPSIA information can be obtained
at www.ICGtesting.com
Printed in the USA
JSHW020148170521
14815JS00006B/8